Counseling in Medical Genetics
Third Edition

Counseling in Medical Genetics
Third Edition

Sheldon Reed

ALAN R. LISS, INC.   •   NEW YORK

Library of Congress Cataloging in Publication Data

Reed, Sheldon Clark, 1910-
   Counseling in medical genetics.

   Bibliography: p.
   Includes index
   1. Genetic counseling. 2. Medical genetics.
I. Title.   [DNLM   1.   Genetic Counseling   QZ50
R326c]
RB155.R36   1980   616'.042   80-22266
ISBN 0-8451-0208-7

Printed in the United States of America

Contents

Preface to the Third Edition vii
Foreword ix
1   "In the Beginning" 1
2   Our Distant Relatives 5
3   A Philosophy for Counseling 9
4   Diagnosis and Genetic Heterogeneity 29
5   The Physician Discovers the Chromosomes
     (and Medical Genetics) 33
6   Amniocentesis and Prenatal Diagnosis 47
7   Treatment for Some Genetic Diseases 55
8   A Few Laws 61
9   The Ubiquitous Heterozygote (or, the Common Carrier) 69
10   Please Don't Marry A Relative (Consanguinity,
     Assortative Mating) 73
11   Blood Genetics 79
12   Disputed Paternity 89
13   Cystic Fibrosis – A Challenge 93
14   Twins 99
15   Multifactorial Inheritance 105
16   Some Common Multifactorial Diseases: Allergies, Cancer, The
     Central Nervous System "Syndrome" (Neural Tube Defects),
     Circulatory Diseases, Clefts of Lip and Palate, Clubfoot,
     Congenital Hip Disease, Convulsive Seizures, Diabetes,
     Duodenal Ulcer, Psoriasis, Pyloric Stenosis 111
17   Normal Traits: Height and Weight, Skin Color, Intelligence 169
18   Mental Retardation 181
19   The Psychoses 187
20   The Environment: Radiation and Nuclear Power, the Drug
     Scene, Fetal Alcohol Syndrome, Infections (Rubella, etc.) 197
21   The Rare Genetic Traits 213
22   Putting the Puzzle Together 219
Literature Cited 225
Index 241

Preface to the Third Edition

The first edition of this book (1955) was written as an introduction for physicians to the new subject of counseling in medical genetics. It was my hope that it would have a wide distribution, and it did. Thousands of physicians enjoyed the comic bits in it, hopefully some learned a little genetics, and all were introduced to genetic counseling.

The second edition (1963) heralded the astonishing discoveries resulting from the new techniques of human chromosome studies and their relationship to an almost endless list of anomalies. Counseling in medical genetics had become established by 1963 and was no longer a novelty. That edition was also published as a paperback entitled, *Parenthood and Heredity*, in 1964 by John Wiley and Sons, Inc.

This third edition comes 25 years after the first one and represents the fruition and maturity of counseling in medical genetics. Counseling centers, and their satellites, are available around the world. The general public has a fair comprehension of what genetic counseling is, and requests it with ever increasing frequency. There is no longer a pressing need to sell the concept of counseling in medical genetics, as the patients will demand the service. The aim of this edition is to provide the information and references that the physician and any other genetic counselor may need in order to provide genetic counseling of satisfactory quality. The tidal wave of new techniques and new information is still rushing in, and some of the material in this edition will probably be obsolete by the date of publication. In the meantime, the general rules and principles should not change much and should be adaptable to the wealth of new information. Due to the lack of availability to physicians of foreign publications in genetics, an effort has been made to include the necessary material in easily obtainable, English language references.

The general public confuses genetic counseling with genetic services and genetic engineering, which are different areas of interest and involve speculations and ethical problems that are independent of genetic counseling. This book should help to separate fact from fancy. Genetic counseling becomes increasingly scientific as the years pass and, fortunately, also more exciting and important as a helping hand for all people. It is the most basic of the various genetic services.

It is a great pleasure to thank Charles M. Woolf, PhD, for reading some of the sections of this book and for his gracious hospitality during the writing of those chapters. I am indebted to V. Elving Anderson, PhD, and especially to Robert J. Desnick, MD, PhD, who read every chapter and helped me with decisions about the content of the book. The final decisions, as well as the errors of omission and commission, are mine.

Mrs. Bertha Storts had the infinite patience to continue the retyping of the text as, almost daily, the endless flow of new information in the journals and books altered the picture of human genetics.

Sheldon Reed, PhD

Foreword

In 1947, Sheldon Reed coined the term "genetic counseling." He defined the primary function of genetic counseling as the process of "providing people with an understanding of the genetic problems they have in their families." He pointed out that "a most important requirement of the counselor was to have a deep respect for the sensitivities, attitudes and reactions of the patient." His pioneering experience in the application of these concepts to the counseling of families with birth defects and his desire to acquaint physicians with the principles of genetics as applied to (the practice of) medicine led him to write the first edition of this book in 1955. It was the first treatise on genetic counseling.

"Counseling in Medical Genetics" was well received. Sheldon Reed's clear, understandable explanation of complicated genetic concepts, his focus on common genetic problems, and his gift of wit interspersed throught the text, made the book both readable and enjoyable. The second edition, published in 1963, provided an update of the rapid developments in medical genetics including an important breakthrough in human cytogenetics (eg, the chromosomal abberations of Down, Klinefelter, and Turner syndromes). The second edition was as popular as the first, was translated into several languages, and has served as a genetics primer for many physicians.

This, the third edition, incorporates the most recent advances in genetic medicine. These sophisticated new developments are described with the same clarity and wit as in the previous volumes. Reed's years of experience and insight into genetic counseling are everywhere apparent. His focus on common genetic problems and the more common birth defects enhances the utility of this text for the practicing physician and genetic counselor. Most outstanding and not to be found elsewhere, this edition provides an up-to-date compilation of the genetics of many common birth defects. The majority of these disorders are inherited as multifactorial traits and Reed's presentation of multifactorial inheritance is one of the highlights of the book. The lucid discussion of these disorders and their recurrence risks, as well as those due to single gene or chromosomal abnormalities, make this a most useful book for practitioners interested in the genetic aspects of medicine. Clearly, this edition of Reed's "Counseling in Medical Genetics" will be valued by both practitioners and geneticists alike.

Robert J. Desnick, MD, PhD

1
"In the Beginning"

The Bible starts with the above three words and proceeds to describe the creation of the world and the evolution of plant and animal life, including man, in the first six paragraphs. Such brevity can hardly be equaled, but an attempt will be made to outline the development of genetic counseling and heredity clinics in as few words as possible.

The fact of heredity was clearly stated in various books of the Bible, and the concept that both the good and bad characteristics of an individual are in large part a biologic legacy from his ancestors was perhaps then more explicitly accepted than it is now. But the precise rules by which heredity works were not known in the beginning; they were first demonstrated by an Austrian monk, Gregor Mendel. It was not until our own century that the significance of the Mendelian rules was appreciated. One of the first to see the applications of the Mendelian rules to the welfare of man was Dr. C. B. Davenport, who established the Eugenics Record Office at Cold Spring Harbor, New York, in 1910.

On March 1, 1927, Charles F. Dight, M.D., wrote his bequest providing funds for the eventual founding of a counseling center at the University of Minnesota. Since Dr. Dight was exceedingly sound in mind and body, it was not until 1941 that the Dight Institute was opened with Dr. C. P. Oliver as director. Meanwhile, the Heredity Clinic of the University of Michigan was initated through the efforts of Dr. L. R. Dice. Several geneticists had given counseling for many years before 1940, but only since then have heredity clinics been recognized as useful institutions in the community.

I began using the term "genetic counseling" between August and December of 1947, and Bulletin Number 6 of the Dight Institute entitled, "Reactivation of the Dight Institute, 1947–1949," and "Counseling in Human Genetics" was published in 1949 by the University of Minnesota Press.

In the first editor of this book, *Counseling in Medical Genetics* (1955), it was stated that, "There are now about 20 places which function as heredity clinics and all give counseling and information free of charge." Since then the growth and proliferation of heredity clinics has been explosive. The fifth edition of the *International Directory of Genetic Services,* compiled by Lynch et al [1977], lists about 400 persons or places where genetic counseling can be obtained in the United States in addition to several hundred in the rest of the world. It is gratifying that this service has been accepted so enthusiastically by the general public.

The policy of the heredity clinics of giving free consultation means that their financial condition is likely to be unstable. Most of them survive only because of the broad-minded generosity of various public and private agencies. Early support for the Dight Institute, in addition to its original endowment, came from generous gifts from the Rockefeller Foundation. The purpose of the Rockefeller grants was to support the counseling program as a research problem, with the expectation that the potentialities of heredity clinics would be determined. The continuing establishment of new genetic service units throughout the world leaves no doubt that genetic counseling is now an accepted part of most cultures.

The reader may appreciate a quick introduction to Dr. C. F. Dight, whose bequest established the heredity clinic at Minnesota and ensures its continuity, even though the major support for the counseling program must be obtained elsewhere.

Dr. Dight had many idiosyncrasies. For some years he lived in a house that he had built in a tree. It was on stilts and was entered by means of a spiral iron stairway. There were appropriate proverbs painted over the doors such as, "Truth Shall Triumph, Justice Shall Be Law." He was the medical examiner for a small insurance company but did not bother with private practice. His money accumulated as a result of exaggerated frugality, shrewd investments, and a calculated failure to file income tax returns. He always had a petition in his pocket.

Unusual people like Dr. Dight often contribute invaluable gifts to society. Dr. Dight was primarily responsible for the adoption of a Minneapolis city ordinance enforcing milk pasteurization. Sixty years ago this was considered a radical attack on the rights of milk companies, but it prevented many cases of undulant fever and other milk-borne diseases. He fought for and obtained an efficient garbage removal service and instigated the foundation of a public market. But his greatest interest centered in the application of Mendel's laws of heredity to the welfare of mankind.

Dr. Dight built his house in a tree because he was afraid of grass fires, but he was not afraid to fight for the use of science and intelligence in the improvement of our social and biologic inheritance. He realized that practically every family had problems resulting from their particular heredity and that many of the problems could be solved if there were a center where the family could get the facts about human genetics. The idea became a fact, and well over 3,500 families or individuals have received education and consequent understanding of problems due to their heredity at the Dight Institute for Human Genetics of the University of Minnesota.

The first legislative declaration to affirm that Dr. Dight was right came in 1959, when the state legislature of Minnesota passed an act authorizing the State Board of Health to conduct a "Program for Study of Human Genetics Problems." A unit for human genetics was established promptly in the Division for Special Services of the State Board of Health. It provides genetic counseling and general education in medical genetics. Both the Dight Institute and the State Board of Health unit are encouraged and strengthened by a group of interested laymen known as the Minnesota Human Genetics League. This vigorous organization was formed according to the terms of Dr. Dight's bequest. It is remarkable that as a result of this small bequest of money, accompanied by appropriate directions, Minneosta became the first state officially to establish a unit for human genetics in its public health program.

Counseling in medical genetics is a most important practical application of the findings of the science of human genetics. It could help almost every family, if available to them. We are still in the beginning stages of the development of sound practices in genetic counseling. As long as we do not take ourselves too seriously, but instead approach the problems in a lighthearted manner, there will be no danger of the gory excesses committed in the name of eugenics in the past. The persons responsible for promoting those outrages were invariably lacking in a sense of humor, a sense of perspective, and an understanding of the human condition.

There is no need to go into the history of genetic counseling here, and there is no definitive work on the subject, though my short essay [Reed, 1974] presents some of the landmarks in the progress of this useful service in the United States.

Finally, a broad definition of counseling in medical genetics may be useful. This comes from a report to the American Society of Human Genetics by an ad hoc committee and was published in the *American Journal of Human Genetics* (27:240–242, 1975). It is as follows:

> "Genetic counseling is a communication process which deals with the human problems associated with the occurrence, or the risk of occurrence, of a genetic disorder in a family. This process involves an attempt by one or more appropriately trained persons to help the individual or family to (1) comprehend the medical facts, including the diagnosis, probable course of the disorder, and the available management; (2) appreciate the way heredity contributes to the disorder, and the risk of recurrence in specified relatives; (3) understand the alternatives for dealing with the risk of recurrence; (4) choose the course of action which seems to them appropriate in view of their risk, their family goals, and their ethical and religious standards, and to act in accordance with that decision; and (5) to make the best possible adjustment to the disorder in an affected family member and/or to the risk of recurrence of that disorder."

2
Our Distant Relatives

The geneticist lives on people's relatives. The science of genetics is the study of the transmission of traits from one generation to another; therefore, the geneticist has to know something about parents and offspring and often studies other relatives as well. But why should the geneticist be interested in our really distant relatives, such as the other primates?

Biologists know that biology makes sense only when viewed in the light of evolution. It is not possible to trace our ancestry back to the earliest mammals, but we are mammals also and descended from them. Our mammalian ancestors of millions of years ago must have had at least a few genes that are still dividing faithfully in our own cells.

It is not surprising that medical research is carried out on mice, cats, and dogs with the expectation that, because they are mammals, the research will be applicable to people — and often is! But these domesticated mammals are very distant relatives indeed. Much closer, though still very distant relatives, are those in our own order of mammals, the primates.

There are no known hybrids between man and any other species. This is not due to moral scruples alone, but rather to the physical unavailability of any species closely enough related to man to produce viable young. However, human cells have been fused with plant cells, an interkingdom fusion. Human (HeLa) cells have been fused with tobacco protoplasts. The human nucleus retained its integrity in the tobacco cytoplasm up to six days after fusion. This clever research was described by Jones et al [1976], but it is not the kind of mating that arouses one's prurient interests.

The genera and species of the primates seem to be rather aloof from each other in sexual contacts, and no crosses had been known to occur between the apes or between primate species with different chromosome numbers until the study of Myers and Shafer [1977]. They report that on August 11, 1975, and August 30, 1976, hybrid female apes were born to a female siamang, Symphalongus syndactylus, and a male gibbon, Hylobates moloch. The mother had 50 chromosomes and the father had 44, while the "siabon" hybrids had 47. There was little similarity between the G band patterns of the parents, but the proteins of the parental species were virtually identical and the animals are very similar behaviorally.

The cover of the July 20, 1979 (vol 205), issue of *Science* had a color picture of the first one of the two hybrids, a healthy three-year-old female. This work suggests that chromosomal rearrangements have occurred quite rapidly in these lesser apes and that there exists a greater genetic distance between them than that which distinguishes the great apes from one another and from man.

This is exciting because it suggests that healthy hybrids might be produced between different genera of the "higher" primates now that all of us live rather happily together at zoos and circuses, among other places. Crosses between chimpanzees and gorillas, for example, would have to be done by artificial insemination, if at all, because these two genera have such different life styles.

Dutrillaux and Rethoré [1975] compared the chromosomes of the pygmy chimpanzee *(Pan paniscus)* with those of the ordinary chimpanzee *(Pan troglodytes)* and found them to be very much alike. Twenty pairs seemed to be completely analogous in the two species. Presumably a fetus would be produced and perhaps a viable hybrid could be obtained from a cross of these two chimpanzee species by artificial insemination.

The frontispiece of this book comes from a study by Warburton et al [1973] in which the chromosomes of the chimpanzee *(Pan troglodytes)* were compared with human chromosomes. The banding patterns in the two species are so similar that the homologies seem clear in spite of minor differences. In the photograph the chromosome on the left of each pair is a human chromosome, and the chromosome on the right is presumably the homologous chimpanzee chromosome. The human chromosome 2 corresponds to two chimpanzee chromosomes, numbers 13 and 17, and accounts for the fact that chimpanzees have 24 pairs of chromosomes while people have only 23 pairs. They suggest that the short arm of human chromosome 2 corresponds to the long arm of chimpanzee chromosome 17, whereas the long arm of human chromosome 2 corresponds to all of chimpanzee chromosome 13.

It is thought that the Pongidae (Pongo, Gorrila, Pan) separated from the Hominidae some 25 million years ago. Probably chimpanzees, gorillas, and man have been drifting apart for from six to twelve million years. Nonetheless, the chromosome banding patterns have remarkable similarity. Chromosomally speaking, one might expect that it would be possible to have Pan-Homo hybrids produced by artificial insemination, as the chromosome differences are much less obvious than those found in the siabon hybrid described above.

Interest in possible hybrids between Homo and Pan is not frivolous. A vast amount of work has been done on the biochemistry, anatomy, physiology, behavior, and ecology of the two species. The genetics of the hybrid would be most instructive for us from many points of view. We know already that the genotypes of the human and the chimpanzee are as similar as those of sibling species of other organisms. King and Wilson [1975] discuss the paradox of why two species with such similar genes differ so substantially in anatomy and way of life.

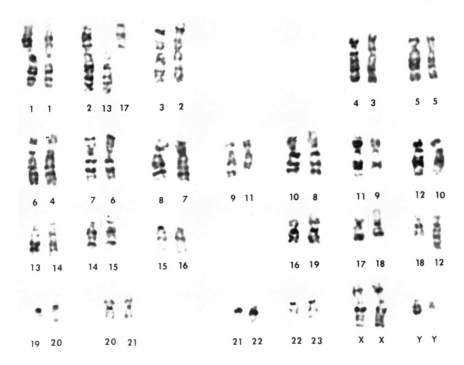

1 1	2 13 17	3 2					4 3	5 5			
6 4	7 6	8 7	9 11	10 8	11 9	12 10					
13 14	14 15	15 16		16 19	17 18	18 12					
19 20	20 21		21 22	22 23	X X	Y Y					

What Is It?

   Is it just another chromosomal aberration? No, the photograph (courtesy of Dr. Dorothy Warburton) is a comparison of human and chimpanzee karyotypes, an imaginary hybrid, if you will. The chromosome on the left of each pair is a human chromsome while the one on the right is from a chimpanzee. Human chromosome number 2 has matching bands in chimpanzee chromosomes 13 and 17. What does this have to do with genetic counseling? See Chapter 2 for some answers.

They propose that regulatory mutations account for the major biological differences between humans and chimpanzees.

A paper by Miller [1977] provides evidence that the gorilla is more closely related to man than the chimpanzee and that all three enjoyed an ancestral type in common many million years ago. Thus the possibility of a human-gorilla hybrid seems more interesting but perhaps less practical to obtain than a chimpanzee-human hybrid fetus. Even fetuses between chimpanzees and gorillas would be of extraordinary interest and scientific value, if they could be obtained.

   Hybrids among the primates would not only provide specific data as to the inheritance of fundamental traits but would also give us a much better feeling for the relative influences of genetic and environmental factors affecting behavioral traits that are so difficult for the counselor in medical genetics to deal with.

3
A Philosophy for Counseling

My philosophy of genetic counseling has already been hinted at in the beginning chapter. Physicians and others should not do genetic counseling unless they have a sensitive feeling for the human condition. This involves a willingness to commune with anyone and everyone at their own level, not at your level. The 'hand-me-a-scalpel' type of authoritarian physician will not be a "good" genetic counselor. The greatest single requirement is the willingness to listen to the counselee, even until the point of pain.

The above is related to the philosophy *of* genetic counseling, that is, the kind of mind-set that the counselor ought to possess. A philosophy *for* genetic counseling is concerned with the ethical, legal, and even political, aspects of counseling and is of great interest to academic philosophers; it does not have a large impact on the daily routine of counseling, except as it is a part of the counselor's cultural background. However, it would be foolish to attempt to separate the academic and practical aspects of any philosophy *for* or *of* genetic counseling.

The "official" 1975 definition of counseling in medical genetics was given at the end of Chapter 1. My personal definition is much simpler: "It is a kind of social work which is often medical but not always so." Of course, it doesn't have to be done by a social worker, but it does have to be done in the way social workers work; that is, with compassion and solicitude for the counselees. See the book by Hsia et al [1979] for an elegant discussion of all aspects of counseling.

Information about heredity, often incorrect, has been provided for families with abnormal children since the development of communication. The advice of relatives, friends, enemies, and neighbors probably has been quite effective in altering the reproductive behavior of the family concerned. The most frequent ideas have been fortified by taboo status or even formal incorporation into religious dogma and civil law. A good number of states and nations have laws regulating consanguineous marriages and most other reproductive behavior, which were taken directly from the Bible. Modern attitudes are greatly influenced by the collections of fact and fancy concerning human reproduction and heredity to be found in all sorts of religious writings. Many of these ancient declarations have been taken more literally than intended by their authors. Others have been so distorted that their modern versions do not have the validity of the original conceptions. Someone has stated our problem in relation to concepts of the past as a prescription to, "keep the embers of tradition but be sure to discard the ashes."

Since advice about heredity is certain to be given by someone, it would seem that the physician or the professional geneticist is in a sounder position to give it than even one's best friend.

The primary function of counseling is to provide people with an understanding of the genetic problems in their families. It seems that almost every family has some troublesome situation directly related to the heredity of one or more of its members. Often the counselor can help alleviate the difficulties.

There may be quarreling between husband and wife as to the "blame" for an abnormality that has appeared in their child. A more dangerous situation exists when the resentment is present but is not expressed. The counselor is almost certain to be of help in these cases.

There may be a sense of shame due to the fact that hereditary diseases often carry social stigmata. Where recessive inheritance is involved, it may be very helpful to point out that carrying a pathologic hereditary unit that is concealed by the normal gene does not mean that the carrier is defective himself. All of us probably carry several such hidden defects, and it is just bad luck when both parents happen to carry the same defect and produce a child showing the disease in the full-blown state. The patient will regain composure when he comprehends that hidden recessive genes are present, not only in his own germ plasm, but in the germ plasm of his friends, enemies, and neighbors as well.

Maternal guilt is an emotional reaction that should be watched for, particularly when the child has a congenital defect. Sometimes the mother may not have wanted to have the baby. She may have made an unsuccessful attempt to abort it. Consequently, she may think that what is really an hereditary defect was, instead, the result of her unsuccessful abortion attempt. Instruction as to the actual origin of the defect should remove, or at least alleviate, her mental anguish.

Explaining Risk Factors

The most frequent and most important function of the counselor is that of stating the chance of reappearance of an abnormality in each child subsequent to the affected one. The patient pays the physician for this information and is not getting his money's worth if he is only assured that, "lightning never strikes twice in the same spot." In the families that came to us after the lightning had struck twice, the mistaken physician was no longer considered to be a family friend.

It has been our experience that one can explain to the parents what the chances are of another abnormal child and that the parents adjust to the facts very well. Being forewarned, they are psychologically forearmed if the next child does prove to have the abnormality. If the child is normal, they experience gratitude and enjoy a mental uplift. In some of the cases that have come to my attention, cases in which the parents had not been conditioned to the possibility of a second abnormal child before it appeared, the mothers' mental processes were badly shaken. While we have helped them to reorganize their psychologic shambles, it would have been better to have prevented the damage in the first place.

Our clients come to us because they are troubled. They show great affection for their abnormal child and give it more than its ordinary share of attention, but the parents are unhappy both for the defective child and for themselves. We have never seen parents who wished to repeat their misfortune. However, the desire to compensate for the loss of a child by the production of a normal baby is often very strong. Thus, they want to know what the chances are of another abnormality. We give them the figure if we have a reliable one; otherwise, we tell them that we do not know the value. The parents often ask us directly whether they should have more children. This question is one that we do not answer because we cannot. The counselor has not experienced the emotional impact of this problem, nor is he intimately acquainted with their environment. We try to explain thoroughly what the genetic situation is, but the decision must be a personal one between the husband and wife, and theirs alone. Of course, if it is possible to help them, it should be done. Perhaps the most useful endeavor is that of merely explaining what chances are; many people do not understand what one chance in four amounts to, as an order of magnitude.

One can resort to various gimmicks in order to illustrate the magnitude of the risk involved. The simpler the explanation the better. When it comes to risk figures, it is not possible to insult the intelligence of the counselees. Even the most sophisticated counselees express horror when percentage risks are mentioned. They do not seem to be insulted when, for example, a 20% risks is described in terms of one's thumb being the risk of an abnormal child and the remaining four fingers, the normal children. Penny flipping also makes sense but black or white beans in a bottle don't go over very well. High, medium, or low risks seem to be understood pretty well, though the counselor has to determine what the couselees mean when they use such indefinite terms.

The decision the parents make may be either eugenic or dysgenic in regard to the hereditary trait under consideration. If they decide to have no more children, it is a eugenic decision; that is, they will not spread further their defective gene either through affected children or normal carriers. If, as usually the case, the chances of producing another defective child are less than the parents feared, then they may have more children. This latter decision is dysgenic, since they will propagate the defective gene instead of arresting its spread. The overall effect of genetic counseling is to encourage people to have more children than they would have had otherwise, since the chances of having bad luck are less than they had assumed (though usually much greater than the "one in a million" chance that may have been given them elsewhere). While it may seem that counseling is dysgenic in regard to the particular abnormal gene, it should be remembered that those people who are sufficiently concerned about the future to come for counseling have commendable concepts of their obligations as parents. These laudable characteristics should become more widespread.

However, the main thrust of genetic counseling must be family oriented and not based on possible improvement of the human genotype. There are no convincing statistics that permit a conclusion as to whether the net effect of genetic counseling is eugenic, dysgenic, or neutral.

The ethical problems related to genetic counseling are not related to any specific religion. The most zealous opponents of abortions often turn out to be Jewish, Protestant, or atheistic in their religious beliefs. There need be no direct connection between genetic counseling and religious precepts. Those who are competent enough to request genetic counseling are already acquainted with whatever contraceptive techniques are available to members of their religion. It is not the duty of the counselor to make religious conversions. He or she has enough problems trying to clarify the mysteries of meiosis to those who come for genetic counseling.

The people that come to me must make their own reproductive decisions, though they may wish me to help them find a marriage counselor or some other particular kind of physician or health care person. However, the genetic counselor should inform the couple of all reproductive options, and they will decide what is suitable for them.

There is a bright light in counseling, as we can expect those who come for counseling to have a much more pessimistic view of their problem than the situation warrants. A frequent question is, "Can I expect to have any normal children?" Consequently, even the clumsiest counselor has an excellent chance of relieving, to some degree, the deep anxieties of his client. The counselor can also help to wash away the guilt feelings of the parents of the affected child by pointing out that genetic calamities can occur in any family and that the parents are not alone in their problem. It helps to remind the client who is suffering from self-blame or self-pity of the biblical phrase, "He maketh His sun to rise on the evil and on the good, and sendeth rain on the just and on the unjust."

The Mechanics of the Art

The informal and unscientific forms of genetic counseling have been carried out for thousands of years. The more or less scientific practice of the art cannot be more than a century old because Mendel's laws were not public property until the beginning of this century. Only in the last few years has there been an effort to standardize the requirements for an optimal genetic counseling service. In The Report on the Council on Accreditation and Certification of the American Society of Human Genetics, June 1979, it was recommended that an American Board of Medical Genetics (ABMG) be incorporated to provide accreditation of training programs and certification of individuals who provide medical genetic services. Genetic services include many functions in addition to genetic counseling, such as chromosome preparation, amniocentesis, and clinical studies. Genetic counselors are defined as "persons with a post-baccalaureate degree who are qualified to participate and assist in the care of persons with genetically caused and predisposed disorders as counselors and coordinators of services and resources." There are these other categories which are open for certification: clinical laboratory geneticist, PhD medical geneticist, and clinical geneticist. This elaborate hierarchical arrangement should result in some upgrading of the quality of genetic services abailable, but it will not prevent the vast amount of genetic misinformation provided by the public at large and by such physicians as are without any competence in genetics.

The genetic counselor cannot do genetic counseling until he has a diagnosis to start with. There is no requirement that he be able to make the diagnosis himself, and it would seldom be advisable for him, or her, to do so in a general genetic counseling situation. The obvious reason for this is that no physician is a specialist in everything. Most pediatricians are not specialists in ophthalmology, for example. The know-how for the genetic counselor who takes "cases" every day consists of experience in knowing who can make a (hopefully) correct diagnosis and in referring the prospective counselee to the specialist for the diagnosis. In my experience the actual situation is the reverse of the above. The specialist refers the concerned persons to me for genetic counseling and, frequently, after some prodding sends along a documented diagnosis.

Many of the situations dealt with are poorly defined. These include multiple malformations, unclassified mental retardation, recurrent spontaneous abortions, drug exposure of one or both parents, chromosomal mosaicism and so on. In fact, those cases where a hard diagnosis can be obtained from a specialist are not too frequent and are so "easy" to accommodate that they are a relief when compared with the usual sticky cases just mentioned. Most of this book will be devoted to the sticky cases, as they are the ones with which the family physician may need some help.

Let me quote once more from the Committee guidelines. "Training in either basic genetics or medicine alone is not sufficient to qualify an individual as a genetic counselor. While the training program is not precisely defined in either time or content, as a generalization it might reasonably include for medical graduates board eligibility in a branch of clinical medicine, training for which should include *at least* one and preferably two years within a specialized medical genetics group. For a non-MD, the PhD degree in genetics or a related field and at least 1 and preferably 2 years of training within a specialized medical genetics group."

Costs of Services

Many counseling centers provide genetic counseling free of charge according to the philosophy that it is an educational and public health activity. This should discourage charlatans and other clearly unqualified persons from hanging out their shingles and developing commercial enterprises. The diagnoses and laboratory tests must be paid for but these costs are often absorbed by research project funds. There is always a cost of the counseling, but this is usually borne by the taxpayer or by the counselee's health insurance rather than as a direct charge to the counselee. Where diagnoses, laboratory tests, and the counseling itself are done by the same group of professionals, the counselee may be charged for the whole procedure without itemization of the bill, a very annoying procedure for the counselee.

A survey was carried out by Associate Professor Terry L. Myers, MD, of Creighton University, Omaha, Nebraska, some years ago as to charges for laboratory services, clinical services, and counseling services. The average costs plus or minus the standard deviation for a few laboratory services were $230 ± $60 for karyotype preparation from an amniotic fluid and $140 ± $54 for Giemsa banding from a blood sample. Clinical consultations ran from $25 to $60. Units which charge for genetic counseling, including pedigree construction, asked from $25 to $50 for the service. The prices varied greatly and presumably were related to the amount of time involved. They all included written reports to the counselee and referring physician. It should be remembered that for each full-dress counseling case there are many phone calls and letters requesting information for which it would be economically unfeasible to attempt to collect a fee. On the whole, genetic counselors will be supported by their professional salaries, by hospitals, or from the public purse. Genetic counseling is not an industry.

Risk and Burden

The genetic counselor is a purveyor of genetic risk figures. His greatest problem is that of putting them in perspective and making them understandable. Murphy and Chase [1975] present a neat table (1-1 in their text), which shows that the risk of death per passenger per 1,000-mile U.S. scheduled airplane flight is only 1 in 625,000 – a risk any rational person would accept because it is lower per mile than that for automobile transportation. The risk per inhabitant per year of being bitten by a dog in New York City is 1 in 294, still a trivial problem. The risk of having a major defect at birth is in the neighborhood of 1 in 33, while the chance of an occurrence of cystic fibrosis in a family where it has occurred already is 1 in 4, and the risk of having a child affected with a simple dominant trait, when both parents have it, is 3 in 4. The last risk given is certainly high, though a rare event, while the risk of death from airplane travel per mile is extremely low but happens frequently, world wide.

The genetic counselor deals with high risks on the scale above, but within this purview there is also a wide range of risks. The risk that any newborn baby will develop a specific rare recessive trait may be only 1 in 100,000, but if both members of a couple are carriers of the trait, then the risk for any child of theirs rises to 1 in 4. Clearly, the risk figure cannot be considered by itself, for the burden of the trait may be of more concern to the parents than the risk figure. The 1 in 4 risk for a carrier mother is the same for both color blindness and hemophilia but the burden is strikingly different. The risk of recurrence of spina bifida may be under 1 in 20, but the burden may be frightful and the parents may decide not to have any more children.

Pearn [1973] has published an interesting study of patients' subjective interpretation of risks offered in genetic counseling. He makes the familiar point that the counselees may obtain some comfort and reassurance if the risk of a repetition is stated as 3 normal to 1 affected rather than 1 affected out of 4 children. The risk is the same but the emphasis is on the normal children rather than on the affected. Personality factors such as optimism or pessimism are extremely important in the way risks are accepted or rejected. The past experience with the genetic trait is also of primary importance not only in how the risk figures are interpreted but also as to how the possible burden is viewed. The following quotation from Pearn's paper shows how bizarre the reaction to the possible burden of the trait can be:

> A woman with both of her two children suffering from coeliac disease came for genetic advice about the risks to further children. A risk of approximately 1 in 10 was given. This turned out to be disappointingly *low* for the woman in question, and and it became obvious that she wanted her next child to have coeliac disease! They were intelligent parents and it transpired that their affected children were doing very well on their diet and the whole domestic microenvironment had adapted superbly to the problems inherent in managing coeliac disease. Both parents also followed the diet, so different cooking schedules were not a problem; there was no gluten-containing food in the house to pose a temptation to the affected children. The parents felt, however, that if a noncoeliac child was born the whole domestic equilibrium would be in turmoil, and the physical and mental expectations of the two children already doing very well could be jeopardized.

Leonard et al [1972] were among the first to emphasize that reproductive attitudes may be determined more by the sense of burden imparted by the defect than by knowledge of its precise risk figures. This has been my own observation also. There have been instances where the parents verbally thrust their defective child into my arms for the benefit of science but with the implication that they didn't really want it back.

Comprehension of the risk figures varies with the sample of persons to whom the figures are given. In the Leonard et al [1972] study where clinic patients

(Down syndrome, cystic fibrosis, PKU, and connective tissue disease) were given genetic information, there was imperfect reception of the genetic message on the part of 27 (44%) of the families. Better understanding and retention can be expected where the persons have been referred to a genetic counselor for their specific genetic problem and have kept their appointment with the counselor. The latter situation is an automatic screening device, which selects a sample that is motivated toward an understanding of their problems.

Briard et al [1977] found that 2 out of 5 counselees have a risk of over 10%, 1 in 5 a risk of 1%–9%, and 2 out of 5 a risk of less than 1% for the trait under consideration.

Goals

The philosophy for counseling is related to the goals that the counselor hopes to achieve. Through cumulative experience, the counselor develops goals which he/she perceives as being essential to the counselees' achieving their personal goals. There have been numerous statements of the goals, and the following quotations are from the interesting paper by Headings [1975].

"1. The counselee feels affirmed as being the most significant person involved in the counseling process and as the principal decision-maker concerning the task jointly entered into with the counselor.

2. The counselee acquires an accurate concept of the inherited entity, its biological basis and prognosis, and the psychosocial adjustments which may have occurred as a consequence of the disorder.

3. Misconceptions held by the counselee or in his/her social grouping are corrected.

4. The counselee's anxiety is brought under control. In some instances this is a consequence of correcting misconceptions; in others, it involves a conscious assessment of anxiety-provoking facts and perceiving them as a realistic component of personal experience.

5. The counselee acquires a clear understanding of available options for modifying prognosis and/or preventing intergeneration transmission, as well as the means by which he/she can achieve the desired results from the option chosen.

6. The counselee develops confidence in his/her coping skills, which may be called for in adapting to the biological and psychosocial properties that characterize his/her situation."

Headings also has some words of wisdom about the counselor: "The professional self-image of the counselor requires specific attention. It requires a move from a traditional image of principal actor in the counseling drama to one of coprincipal with the counselee in which there is a shared search for healing and wholeness. That the counselee can experience caring and empathy from the counselor seems to be underestimated by most counselors." What Headings means is that if the counselor has humility and generosity and understands the human condition, he will be personally successful with the counselees and they will sometimes name subsequent children after him or her. My greatest vanity has been in those few cases where a child was named after me. I also received vicarious pleasure when one of these "namesake" babies recently became a national hockey star.

Results of the Philosophy

It is not possible to measure all the results of genetic counseling. We have followed up our "cases" many years after they originated, but we cannot have learned of all of their ramifications. At the time of counseling we are careful *not* to ask the counselee to let us know how things turn out, as this might convey a threat that in their case the results might be unfavorable. They could get the idea that our reasurance was not valid but only an attempt to cheer them up.

A very useful attempt to evaluate the success of genetic counseling was carried out by Antley and Hartlage [1976]. They found that anxiety, hostility, and depression levels were significantly higher in parents of Down syndrome children seeking genetic counseling than in normative controls. Following genetic counseling, there was a significant lowering of anxiety ($P < 0.0005$) and depression ($P < 0.05$) along with a significant increase in overall self-concept ($P < 0.01$). These findings indicate an improvement of the mental state of the parents, which would seem to be the most important goal of counseling. These findings should also be found to a greater or lesser degree for other traits in addition to the Down syndrome.

In my own counseling there are no tests or questions posed for the counselees upon arrival, as that is an excellent way of turning them off. They have been over-interrogated already. There is sufficient information with the referral so that I can talk until they start to ask questions, and from then on the information flows freely and the interaction is candid and comfortable. I am, of course, deprived of any test results with this system and can use only follow-up requests in later years to evaluate the results of the counseling. The most striking aspects of the follow-up are the subsequent reproductive histories, and we are inclined to over empha-size them, even though they represent only one kind of result of the counseling. This fascination with the subsequent reproductive behavior of the counselees is probably in part a holdover from the ancient concern with eugenics. The divorce between genetic counseling and eugenics must still be finalized in its entirety.

Nonetheless, the subsequent reproductive history of the counselees is of interest. Larson and Reed [1975] conducted a follow-up of the Dight Institute cases selected because the couples had a risk of producing a child with a genetic anomaly. Reaction to the information received was favorable with 90.2% of those responding indicating that the information they received was quite helpful or somewhat help-ful. The follow-up was from 4 to 18 years after the counseling.

The counselees were divided into those who had a low risk, under 10%, and those who had a high risk of over 10%. Over 60% of the counselees remembered their risk figure after the 4–18 years. Most of those who did not remember the risk figure correctly were in the low risk group.

We found that 46% of the high risk cases had no subsequent children. In the low risk cases, 33% had no subsequent children. In the high risk group 28% reported sterilizations while only 1.3% of the low risk group had obtained a sterilization. This last figure is lower than the national average for people without genetic problems. Probably some 10%–25% of married couples today obtain a sterilization as a way of completing their families. Whatever the true figures are at present, it is clear that sterilization is a routine part of medical practice and a widely accepted method of permanent contraception.

One of the most important successes of our counseling has been the relieving of needless fears, which are very strong in many cases. Every genetic counselor has encountered clients whose risk was very low but who asked whether they could have any normal children. In an occasional case the counselee asks this question even though a normal child has been produced subsequent to the affected one.

Klein and Wyss [1977] reported on 1,000 genetic consultations and showed that for 306 low-risk couples there were 290 subsequent children and only 8 (2.8%) had the dreaded affection. However, it was striking to note that 19 (6.5%) children were affected with some other unforeseeable anomaly, often of multigenic type (cleft lip, etc.). Their 338 high-risk couples had 83 subsequent children, of whom 15 (18.1%) children manifested the feared disease and 2 (2.4%), another disorder.

The above data show an unpleasant and perhaps unexpected fact that should be borne in mind by genetic counselors. The fact is that even when the risk is low, there is always an additional "background" risk of about 3%–5% that some unexpected trait will appear which may be worse than the one for which the original counseling was obtained. The fact that the second anomaly is genetically independent of the first one doesn't make it any less burdensome. It is probably even more so because the parents are not acquainted with it.

There are now numerous accounts of the results of genetic counseling, one of the most comprehensive is that of Evers-Kiebooms and van den Berghe [1979]. The report of Koch and Schwanitz [1977] is concerned with counseling for chromosome anomalies.

The best evidence of the success of genetic counseling is its explosive growth throughout the world.

Ethics

Philosophers and theologians have been greatly excited and titillated by the ethical component of genetic counseling. Some seem to think that there is a baleful conspiracy behind it, which somehow was designed to destroy the family and our traditions. The basic philosophy of genetic counseling is simply the desire to help the family solve its genetic problem in whatever way is most helpful to it. As in ecumenical churches today, one can either kneel or sit, whichever is comfortable for the person. The counselor does not drive the family car for them, he rides along with them and shares the view.

Such a simple philosophy does not provide much material for analysis by the philosophers, so genetic engineering is dragged in although it has little to do with genetic counseling. There is great interest in cloning, but this is of no relevance to the genetic counselor. We have natural clones of identical twins or triplets in every state of the union but the only question the genetic counselor needs to answer is about the genetics of twinning. I have little interest in cloning because it is obvious that the products would not repeat the performance of the donor of their cells, because the environmental component would be different in all cases. The genetic counselor must not only divorce his subject from eugenics but also from the exciting but unrelated areas such as cloning. These are of great importance to the human geneticist but not to the genetic counselor.

It should be stressed that there are numerous philosophical, ethical, and legal problems related to human and medical genetics, but only a few of these relate to genetic counseling.

There are a few areas where the genetic counselor runs into ethical problems. One of these is involuntary mass screening. Genetic screening without genetic counseling would seem to be useless except for research purposes. Numerous laws have been enacted requiring the screening of black school children for sickle cell anemia. The information obtained could not be of any possible use to a school child or its parents. Such laws were ill-advised and have been repealed; it is doubtful that the advice of any geneticist was requested before such laws were passed. In this case, *affected* children would have been detected by a physician before school age. The intent was to detect the *carriers* of sickle cell anemia, but children about to enter school are not about to reproduce, so the information would not be of practical value to them at that time. However, if black persons wish to find out before reproduction whether both members of the couple are carriers, then the facilities for both testing and genetic counseling for all those tested should be available.

Genetic screening at birth for PKU or chromosome anomalies is quite different from the sickle cell carrier screening just considered. In such cases the detection of the affected child early can result in significant treatment and amelioration of the disadvantages resulting from the biochemical or chromosomal defect. These benefits far outweigh the fact that technically the drops of blood taken from the heel prick or umbilical cord may have been taken without informed consent from the mother. However, this procedure harms no one and helps many so that the objections that the Fourth Amendment right has been violated are foolish

have it. The affected should make the sacrifice rather than the unaffected. In the past the heaviest burden of the disease has been borne by those without the disease, especially the spouses and children of the person with the disorder.

The test is not available yet, other than on an experimental basis. The dilemma is still a theoretical proposition as has been most of the discussion of morals and ethics relating to the genetic counseling area. How does one counsel the young couple that wishes to start a family but have the 50% risk hanging over them for ataxias, Huntington's, and other serious dominant traits? My attitude has been that it is not fair to the 50% who are free of the undesirable gene to be deprived of children. We cannot tell which ones they are, so all must behave in a way which will reduce the tragedy in the next generation to its lowest terms without a complete ban on reproduction. In other words, have a small family, as most Americans do, and try not to exceed three children, with one or two being a better number under the circumstances. The desire to replace oneself is very strong and should be discouraged only when there is proof that the undesired gene or chromosome anomaly is present, not when there is just a statistical probability of it, especially if the probability is small. The risk of 50% is high, and many counselees will consider that risk to be too high for them. For those who consider the 50 percent risk acceptable, the counselor can point out to them that they do not need to enter any baby-producing contests, that it would be extremely unwise to flaunt their virility under these circumstances. A small family should suffice.

There is one more important ethical situation relating to tests for the carriers of X-linked traits with serious consequences such as Duchenne muscular dystrophy, hemophilia, and Lesch-Nyhan disease. Let us assume that tests for the women who carry one of these X-linked traits are accurate and that the 50% of the sisters who are free of the undesirable gene are so notified and are enjoying the release and relaxation from fear that the information provides. The other 50% of the sisters who turn out to be carriers of a gene for one of these serious traits will be in quite a different situation. Half of their sons will be affected regardless of whom they marry. My experience with women who were often extremely perturbed by the 50% probability that they might be carriers of the gene was that they frequently got themselves sterilized without inquiring of me as to whether it would be advisable for them. The reason for their drastic action was the result of their having experienced the agonies caused by having one or more affected uncles or brothers.

The development of amniocentesis provides a different kind of dilemma. Those women who find they are carriers of an X-linked gene of this sort can find out by amniocentesis whether their fetus is a boy or girl. The girl would continue normal development but the male fetus would have a 50% chance of being affected. However, it also has a 50% chance of being perfectly normal. The dilemma then is that one must take an equal chance of destroying a normal male fetus in order to make sure that an affected male is not born. Women who have to make this decision probably will decide, reluctantly, not to have sons. They will have made the decision, in principle, before having embarked on the amniocentesis procedure. For certain X-linked disorders not previously prenatally detectable, there has been another recent advance.

Fetoscopy, the visualization of the fetus during amniocentesis, has been accomplished in a few centers. This technique allows visualization of the placental vessels, and by using an extremely small guage needle (attached to the fetoscope), fetal blood sampling can be accomplished. In this way, fetuses at risk for various homoglobinopathies and X-linked Duchenne muscular dystrophy can be identified. Current efforts are directed toward the prenatal detection of male fetuses at risk for hemophilia by this method. In this way, women carrying a male fetus at risk (50%) for a severe X-linked trait may be able to carry normal males to term instead of electing termination of all male fetuses.

New Horizons for Counseling

Optimum genetic counseling in the future will be provided by a team constituted of physicians, basic scientists, a genetic associate, a social worker, a psychologist, the genetic counselor, and others depending upon the genetic problem. All this retinue should be available to the genetic counselor, but to assemble all of them at one time would be a genetics conference and not the ideal environment for genetic counseling. The counselees do not care to be the center of a circle of experts in white coats. It should be remembered that the counseling must be for the benefit of the counselees and not primarily for the benefit of the medical school.

The steps of genetic counseling, in briefest form, are: 1) insistence on the most accurate diagnosis possible; 2) establishing the relevance and mode of inheritance; 3) exploring the mind sets and problems of the counselees (and the counselor); 4) determining the need for future tests or contacts; 5) discrete probe into concepts about countraception, if it is thought that the counselees are naive in this area; 6) leave the couple with a feeling of friendship, rapport, and an open readiness to meet again if anyone wishes to do so.

With these ideas in mind, what is the future for genetic counseling? Will the barefoot doctor philosophy in mainland China provide genetic counseling services for the one billion people who live there, or is genetic counseling a luxury available only for affluent small countries like Denmark, where it got an early start?

Murray [1978] provided a very thoughtful article entitled, "Genetic counseling: boon or bane?" in which he contemplates the possible future of genetic counseling. One of the banes would be the appearance of a "new" kind of genetic counselor paid by the state to see that persons with defects were not born so they would not be a drain on its resources. There would be no concern for the needs of parents — only for the cost/benefit ratio or boon/bane ratio projected for the individual. He would prefer, and so would I, to see man become extinct in the process of following principles based on love and human concern for the needs of all human beings than to ensure our survival under regimented, inhuman programs in which we are treated as so many computer punch cards. I hope we are being unrealistic as far as the ultimate extremes of the concept are concerned.

The genetic counselor should do his counseling in relative peace and quite, with privacy and confidentiality strictly observed. He cannot do his job without the support of a cytogenetics laboratory, a biochemistry unit, sophisticated diagnostic facilities, and the clinical expertise of numerous specialists. A genetic counseling center must be located at or near a large medical center. As Fraser [1974] has pointed out, it is highly desirable that the counseling services be associated with an ongoing research program. However, the genetic counselor must be where the population is, if the need is to be met. Satellite genetic counseling facilities have been developed to bring the counselor and the counselees into physical contact. Epstein et al [1975] presented an illuminating description of the California center-satellite system.

Again: how large is the need, what is the demand? A study by Trimble and Doughty [1974] included 30,603 liveborn children born in British Columbia from 1952–1972. The persons were followed up because many conditions tend not to get registered until the affected individuals reach school age. The birth cohorts for 1964–1966 seemed to be the most representative of all and therefore were chosen for special consideration. The total for all registered conditions believed to be hereditary, or partially so, is about 6 per 100 liveborn and this is, of course, only for those traits that manifest themselves during childhood or youth. However, the figure does include all congenital malformations, whether a genetic cause has been demonstrated or not, on the belief that a significant genetic contribution exists for the great majority of them. This problem is not important for genetic counseling because all of them were regarded as handicapping, and those of environmental origin would be almost as likely to present themselves for genetic counseling as those where genetics is obviously of importance.

It is astonishing that 6%, or about 1 in 16 births, were registered as handicapped, presumably due to genetic causes! For the world at large this means that millions of births every year have handicaps which might warrant genetic counseling. Genetic counseling is now available for less than 100,000 persons per year. Clearly, genetic counseling, like other important medical services, is unevenly distributed and in no large area is it even adequate. This poor showing is partly due to the fact that the vast majority of the world's population doesn't know that there is such a service. Hopefully, personnel will be available to fill the need as people learn that genetic counseling exists.

The breakdown as to the major causes of the handicapping conditions was as follows:

Single gene anomalies	0.178%
Chromosome anomalies	0.162%
Congenital malformations	3.580%
Other multifactorial	1.582%
Unknown	0.598%
Total	6.100%

These figures do not include any disorders of late onset such as the psychoses, neurological disorders, diabetes and so on. They do not include the major circulatory system diseases, which cause the majority of deaths in the population and that have some sort of genetic component. It is a platitude that all traits have a genetic component and every person displays numerous conspicuous traits. Naturally if the trait is burdensome and has simple gene inheritance, the person or the parents of the person will be more likely to try to get genetic counseling than otherwise.

The conclusion from these considerations is there could be a demand for genetic counseling from practically every family throughout the world. Certainly, the demand will continue to rise as the service becomes known. There is no obvious way in which the eventual total demand for personal genetic counseling can be met. The costs might eventually exceed the benefits, unless genetic counseling became an effective eugenic agent through some scientific advance.

That is, if yet-unknown techniques become available that would transform genes from the undesired form to genes giving normal health — and the genetic counselor directed the process — then society could afford an increase in the number of genetic counselors by several orders of magnitude. This is an appropriate dream, but I have no idea as to the likelihood of its materializing.

There is one concept that has never been sufficiently emphasized and that is the cumulative amount of genetic disease in any population. The frequency of the individual anomaly may be low, say 1 in 10,000 for PKU. Some physicians were opposed to screening for PKU on the grounds that, with such a low frequency, they would never have a case in their life-time practice. This was an irrelevant argument because the purpose of the screening was to find the affected children early enough so that the treatment would save them from lifelong mental retardation, not to provide "cases" for the physician. The cost of the screening is modest and the benefits tremendous. What is ignored is the cumulative frequency of all the genetic diseases put together. Let us assume that there are only 1,000 such diseases, each with a frequency of only 1 in 10,000, then 1 in every 10 persons has a serious genetic disease expressed sometime before death. This estimate of 10 out of every 100 persons with a serious genetic disease is certainly conservative, as we have just seen that 6 out of 100 persons studied by Trimble and Doughty

[1974] demonstrate a severely handicapping genetic condition in childhood. Clearly the prospective demand for genetic counseling is greater than the prospective supply of counselors, and this results from the cumulative effect of many traits, each with a low frequency.

Medical research has resulted in the control of many infectious diseases and in the improvement of medical care. As a result, genetic diseases have increased in their relative importance in the population. Roberts, Chavez, and Court [1970] found in hospital deaths among children that genetic conditions were directly or indirectly involved in over 40% of the cases. These were early deaths and therefore serious enough to require genetic counseling even for some of the environmentally important cases, because there is no labeled "environmental counseling service" known to me. The genetic counselor is expected to know the literature concerning environmental risks from drugs, radiation, lead poisoning, and so on. No other specialist assumes responsibility for both environmental and genetic risks.

We see, once again, the evidence for a potential demand for genetic counseling that will greatly exceed the supply of counselors, particularly as the financial support for genetic counseling has always been trivial, and one could predict that it will be inadequate in the future. How will the problem of a shortage of genetic counselors be resolved? I don't claim to know the answers but can offer the following possibilities:

1. The average family physician will continue to be apathetic about genetic counseling and will be slow to either become competent in the area or to refer his families to genetic counseling units. Thus the demand will increase slowly and new counselors will become available fast enough to meet the demand.

2. If population pressure increases too rapidly, one can expect medical services to deteriorate and genetic counseling would suffer along with other specialties.

3. The optimistic stance would be that population growth can be controlled, science will provide unexpected new techniques that will increase affluence so that greatly improved medical services, including genetic counseling, will become widespread.

This last rosy outlook is perhaps more realistic than I think it to be. We have seen a remarkable increase in interest of women in their bodies and their control of them. Women are increasingly reluctant to be forced to produce souls for heaven or cannon-fodder for future wars. Furthermore, future research in genetic medicine will certainly provide some unexpected advances, which will require genetic counseling as a part of the utilization of the discoveries. Whether or not the manipulations will appeal to society, and therefore be accepted, is quite distinct from genetic counseling. However, genetic counseling may be requested in explaining the genetic risks to the family involved with the new technique.

One may get some notion of how scientific discoveries on the distant horizon may be useful in dealing with undesirable single gene traits. There are many clearly undesirable genes that should be eliminated. If their elimination turns out to be a mistake, they can be expected to return to the gene pool as new mutations occur. Society would be happy to reduce some of its variability due to disastrous genotypes, which are a burden at the present time, and maintain its variability in other genotypes.

Considerations of the quality of children are more controversial than those of quantity. In order to live satisfying lives the best average number of children per person is a little over 1.0, so that couples will have two or three children as a rule. It is the duty of governments to educate the citizenry to accept the concept that the country cannot support unlimited reproduction. If one family has over six children, then two other couples must be satisfied to be childless. The logistics of this concept for the population of a planet of fixed size should be self-evident. The governmental obligation regarding the genetic quality of the population can only be stated as one of a desire for genetic improvement, but this criterion has not yet been defined, and a specific program for genetic improvement is not to be expected on the horizon for a long time.

My view of the future of genetic counseling is that there will be a very large demand for it and that, as a result of the demand, some way will be found to finance the enterprise. What the total dimensions of the edifice will be, eventually, is beyond my comprehension. Our skills in counseling should increase. It is important to remember that genetic counseling is a quiet and personal procedure. It must be divorced from possible governmental eugenics programs, and it must also maintain its independence from genetic engineering or whatever flamboyant genetic speculations may be in vogue at the moment. It must not be unethically exploited by the media or used by politicians for their own ends.

Conflict might develop between genetic counselors (who should have the needs of the individual family as their primary loyalty) and politicians (scientific or not), who would promote social programs of genetic control or other "future shock" types of governmental regulation – or even repression.

Widespread screening for genetic traits may turn out to have a very favorable cost-benefit ratio for many genes. However, screening without adequate genetic counseling is not tolerable. Governmental regulation may be necessary for screening programs, but it must be humane or it will be objectionable and inappropriate.

Widespread screening produces dilemmas unrelated to genetic counseling, but of interest to the general public. Should parents refuse to consent to the PKU test for their child, their refusal for whatever motive would be a type of child neglect. Should a child be shown to have PKU, the courts would be justified in forcing the parents to provide the low phenylalanine diet for the child to save it from mental retardation.

The new horizons for genetic counseling seem to be indistinct at this time, but the trends toward growth of the service and better quality counseling do seem to be entirely reasonable expectations.

4
Diagnosis and Genetic Heterogeneity

The best geneticist in the world cannot give correct genetic counseling with a wrong trait diagnosis. Genetic counseling for one recessive trait may be much like that for a different recessive trait, as far as the 25% risk is concerned. However, one cannot depend upon such good luck very often, and it is imperative that the correct diagnosis of the trait be at hand in order to provide satisfactory genetic counseling. The most fundamental error one can make is to assume that a diagnosis provided by the counselee is correct. It may not be. Consequently, it may be necessary to insist upon a rediagnosis. The genetic counselor must have had some clinical experience in order to make a decision as to whether a rediagnosis is necessary. I learned this fact of life in genetic counseling early.

Dwarfism

The most striking case of the disaster resulting from faulty diagnosis in my experience was the following, which occurred some 25 years ago. A mother wrote that her first child had been diagnosed as an achondroplastic dwarf by a physician. She wanted to know what the chances were of a repetition of this trait. My mistake was in assuming the diagnosis was correct and that the affected child represented a new mutation as neither parent was affected. My reply suggested a low risk, as would be expected for a new dominant mutation in one of the gonads of one of the parents. A few years later I received a note stating that she had produced two more children, both achondroplastic dwarfs. This made no sense, so I wrote for a description of the three affected children. There wasn't any question but that they had been misdiagnosed by two different physicians and were cases of recessive diastrophic dwarfism. The risk of a repetition was therefore the 25% expected for a Mendelian recessive and not the small risk expected for a new dominant mutation to achondroplastic dwardism. Consequently, I have been suspicious of all diagnoses since that time. Diagnosis is the soft underbelly of medical practice, and the genetic counselor must always be on guard in order to reduce errors due to wrong diagnoses.

Dihydropteridine Reductase Deficiency

Mistaken diagnoses are not a relic of the past but continue into the present. The article by Kaufman et al [1975] is an interesting account of a case of dihydroperidine reductase deficiency that was misdiagnosed as classical phenylketonuria. This error had severe medical consequences, though the genetic counseling risk would have been the same (25%). The parents of the patient were first cousins.

The diagnosis of a trait may be easy in some cases even though there may be great variation in expression of the trait from patient to patient. Such variability will often be due to genetic modifiers and environmental differences. The constellation of small genetic and environmental modifiers must be enormous. One can hope, at best, to identify some of the major factors involved. The severity of a particular trait may vary from one group of family members to another. One kinship may have early onset of a trait while in another kinship the affected members will display a late onset.

Difficulties With Down Syndrome

There may be problems with traits where the diagnosis is uncomplicated, as with the Down syndrome. Here, one of the parents could be a chromosomal mosaic for the trait with an increased risk of a repetition of the trait in a subsequent child. One of the parents could be a carrier of a chromosomal translocation with a substantially increased risk, about 10%, of a repetition. The most usual case would be that the Down child resulted from chromosomal nondisjunction, with only a small risk in the neighborhood of 1% of a repetition of the syndrome in a sibling.

The problem of obtaining a correct diagnosis is thus a difficult one from many points of view, and the genetic counselor should not shirk his duty to search out the specialist who can provide the best diagnostic service in order that the genetic counseling be based upon scientific fact and not fancy.

The diagnosis and genetic heterogeneity may be related to each other in a practical way. For instance, the same bony alterations are found in autosomal recessive Hurler syndrome as in the X-linked Hunter syndrome. However, if corneal clouding is present, we have Hurler but not Hunter. Other diagnostic differences give a clear distinction between the two superficially similar appearing syndromes. Genetic counseling will be different for an autosomal compared with an X-linked trait, so the correct diagnosis resolves the genetic heterogeneity very nicely.

Dr. Huntington recognized "head bobbers" in one of his Huntington's chorea families as different from the behavior in his other choreic families. Probably there was more than one dominant mutation for the trait among his families; these dominant mutations may have occurred independently in different "founding fathers." I am unaware of any Huntington disease cases where any Mendelian mechanism other than dominance was present. Consequently, genetic counseling can be done with some confidence that a Mendelian dominant gene is involved.

The situation is quite different for numerous other traits where an apparently homogeneous trait depends upon different Mendelian mechanisms for its transmission. I am aware of at least five variants of the Ehlers-Danlos syndrome (hyperelastic skin), three of which have autosomal dominant inheritance. The fourth and fifth types are autosomal recessive and X-linked recessive; probably there are other genetically distinct types as well.

Albinism

Albinism is likewise an interesting diagnostically and genetically heterogeneous trait. Here the different gene loci involved are all recessives, except for ocular albinism, which is X-linked in some families and autosomal recessive in others. According to Witkop et al [1978] there are at least six oculocutaneous forms of albinism (OCA), which are inherited as autosomal recessive traits. I presume they are at six different gene loci but not necessarily on six different pairs of chromosomes. These are tyrosinase-negative OCA, tyrosinase-positive OCA, yellow mutant OCA, Hermansky-Pudlak syndrome (HPS), Cross syndrome (CS), and Chédiak-Higashi syndrome (CHS). They can be distinguished from one another on various clinical and/or biochemical criteria.

Tyrosinase-negative and tyrosinase-positive albinos resemble each other more than any of the other types and the two tyrosinase types could be confused unless diagnosed by a specialist, though there is a little melanin in the skin, hair, and eyes of the tyrosinase-positive person. The similarity of these two types of albinism leads to possible counseling problems. Sight-saving schools and other social functions bring albinos together, and propinquity is essential to reproduction. Should a tyrosinase-positive person produce a child by a tryosinase-negative person, we would expect a normal appearing child — if the two types are dependent upon different gene loci. Albino children would be produced if the gene for tyrosinase-positive and that for tyrosinase-negative were alleles at the same locus. The double heterozygotes from a union of a tyrosinase-positive and a tyrosinase-negative person are normal in appearance. This indicates independent gene loci for those two types of albinism even though the number of such unions is extremely small.

Let us emphasize again that the correct diagnosis is an absolute necessity for proper genetic counseling, even though it is distinct from the counseling process itself. The counselee frequently does not know the scientific name of the trait involved and often shows little interest in learning it. The coulselor's first duty is to make sure that the best possible diagnosis of the trait has been obtained and that he is not being led astray by the various aspects of possible genetic heterogeneity.

5
The Physician Discovers the Chromosomes
(and Medical Genetics)

The first edition of this book *(Counseling in Medical Genetics – 1955)* was truly a pioneering effort because the subject of medical genetics hardly existed at the time. It is difficult to believe today that it wasn't until the next year that Tijo and Levan [1956] clearly demonstrated that the normal number of human chromosomes is 46. Not until 1959 was there proof that a human malady (Down syndrome or mongolism) was the direct result of a chromosome anomaly, shown to be the case by Lejeune et al [1959].

A tepid interest in medical genetics had developed earlier due to the curiosity of the medical world as to whether the irradiation from an atom bomb would cause gene mutations. The possibility that physicians' own x-ray machines might cause gene mutations was considered to be remote. Human genetics and its specialty, medical genetics, was considered to be an academic subject without much clinical significance by most physicians. However, since about 1960 the explosive growth of human biochemical genetics and human chromosome study have created a small revolution in human genetics, which is known as medical genetics.

The chromosomes can be seen and studied with various techniques. Seeing is believing. The practicing physician not only discovered human chromosomes but medical genetics along with them. Down syndrome was no longer a mystery; the three members of the 21st pair of chromosomes responsible for aberrant development of the fetus could be seen and photographed. The cause, the extra chromosome, became very real. The flood of information linking birth defects to chromosome aberrations has not yet subsided.

Any genetic disease is the result of a qualitative difference in the two members of a gene pair or a quantitative difference in the number of members at a gene locus. The gene for albinism and its normal partner are qualitatively different, whereas in the Down trisomy, with three members of each gene on the 21st chromosome in each cell, the difference is quantitative. The child with this trisomy is not lacking genes nor are his genes defective; he has 50% too many of the genes located in the 21st chromosome. Too much or too little chromosomal material in an individual's cells leads to clinical disorders. There are now more than 100 recognized chromosomal disorders, and clinicians in all specialties need to know something of the general characteristics of them, as well as those in their own area of practice. We are fortunate to have a comprehensive listing of the chromosome variations in man compiled by Borgaonkar [1977], 2nd edition. We are fortunate also to have the *Clinical Atlas of Human Chromosomes* by de Grouchy and Turleau [1977]. This book devotes a chapter to each chromosome pair and its syndromes.

I am grateful to Robert Desnick, PhD, MD, for the photographs of human chromosomes provided in this chapter. Figure 1 shows the normal female with two X chromosomes; Figure 2 is the karyotype of a normal male with an X and a Y chromosome, and Figure 3 shows the chromosomes of a male with Down syndrome having an extra chromosome of the 21st pair. It will not be possible to consider the entire mass of chromosome anomalies in this book. The physicians shown obtain a thorough explanation of the particular chromosomal abnormality of concern from the person responsible for making the preparation.

Polyploidy

If there were three members of each chromosome pair instead of the standard two members, we would have a polyploid person. In this case the person would be a triploid, and one can envision also a tetraploid in which the person has four members of each chromosome pair. One might expect such persons to be normal, as with equal numbers of chromosomes of each type they should be in "balance" with no excess or deficiency of a specific chromosome. However, this is not the case.

The reports of Walker et al [1973], Wertelecki et al [1976], and Jacobs et al [1978] stated that triploid infants usually miscarry or survive for a few hours only and they show a variety of abnormalities. These abnormalities are of the type frequently found with the usual chromosomal aberrations such as low set ears, coloboma of the iris, the single palmar crease, and congenital heart defects. The most common anomaly is hydatidiform degeneration of a large placenta. Triploids can have the XXY, XYY, and XXX genotypes but, including all of these, the Walker report [1973] lists only 13 that survived as long as 24 weeks gestational age. There is no explanation as to why the survival of completely triploid fetuses is so poor. Early spontaneous abortion might explain their extremely low frequency. Fetuses with complete tetraploidy, that is, four members of each chromosome pair, are even more rare than those with complete triploidy.

Sex Chromosome Anomalies (Aneuploidy)

The picture changes completely when girls with trisomy for the X chromosome, but with the remaining chromosomes in normal pairs, is considered. Tennes et al

[1975] screened 38,000 births between 1964 and 1974. They found 69 children with sex chromosome anomalies, and 12 of these had the 47,XXX karyotype, that is, trisomy for the X chromosome. The literature includes some 200 individuals with trisomy X, but the report just mentioned follows the development of 11 girls with the 47,XXX who were followed from birth and cytologic identification. About 1 per 1,000 newborn girls have trisomy X, which makes the trait practically a "common" disorder. The parents of the 11 girls were advised of an unusual finding in the cells of their child and invited to participate in the research. The genetic counseling was directed toward maximizing exchange of information and minimizing arousal of anxiety about the chromosomal anomaly. Nonetheless, the information did arouse concern among the parents, as otherwise they would have suspected nothing at the time of birth. The affected girls and their normal controls were examined quarterly for the first two years and biannually thereafter. Delay in early motor development and speech, a mild intellectual deficit, and disturbance in interpersonal relationships occurred in about one-third of the 11 girls, while the other two-thirds were considered to be normal and adequately adjusted. No consistent phenotype or syndrome was delineated.

The importance of the 47,XXX females for genetic counseling rests upon the fact that there may be as many as 200,000 of them in the United States, see Stine [1977, p 153]. Most all of them probably have been accepted as normal or variant members of the general population. What has been their reproductive performance? Statistically speaking, we would expect half the egg cells of the XXX woman to have only one X present and the other half to have two X chromosomes. The latter half of the eggs with two X chromosomes would result, upon fertilization, in 47,XXX females and 47,XXY (Klinefelter type) males. The option of avoiding the birth of these two types of anomalies is available with the aid of amniocentesis if the 47,XXX mother has been identified by chromosome studies. One can speculate that the 47,XXX mother may have no objections to producing a child like herself in some cases, but one can also guess that she would not look forward to the birth of an XXY son with complacency. It is not the function of the counselor to speculate about the choices of the mother but rather to do his compassionate best to provide her, and her spouse, with a complete understanding of their situation. It is surprising that so few cases of 47,XXX females have had their reproductive performance studied considering the fact that there are perhaps as many as 200,000 of them in the United States alone. More work remains to be done with them, particularly concerning their fertility.

A little space should be given to what is called Turner's syndrome. Persons with this anomaly have only one X chromosome, or 45, XO. This is a unique syndrome in that it is the only viable monosomic. Fetuses with only one member of any of the other pairs of chromosomes do not survive until birth. The Turner syndrome person is a sterile female less than five feet tall at maturity with webbing of the neck and other small anomalies, but with normal intelligence in most cases. The

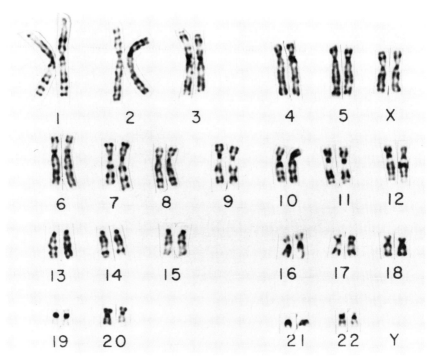

Fig. 1. The chromosome complement of a normal female, 46 XX.

frequency of the syndrome is about 1 in 2,500 females, so that there are about 85,000 of them in the United States. They are of less interest for genetic counseling as they do not produce offspring and therefore do not have to make decisions about the risk of a repetition of the anomaly. See Figure 4 for the karyotype of a person with the Turner syndrome.

Persons with the Klinefelter syndrome are 47,XXY. They are always male by virtue of the Y chromosome, but they have variable male genital development. The number of such persons is relatively high in that 1 out of every 400 live male births has the extra chromosome, or about 530,000 affected persons in the United States (see Stine [1977, p 153]). Of the high-incidence chromosomal disorders, Klinefelter's is one of the most amenable to treatment. Fortunately, they usually have normal sized penises, but they are usually sterile and thus do not have genetic counseling type decisions to make. They need education and treatment, but ordinarily this would be done by someone other than a genetic counselor, though they sometimes come to talk with the counselor (see Fig 5).

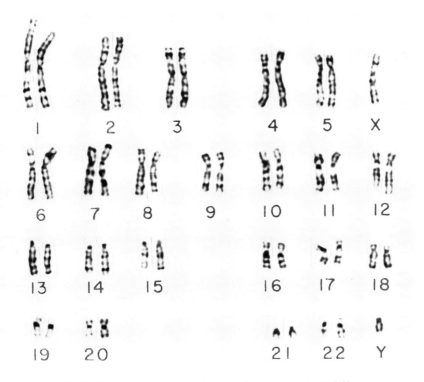

Fig. 2. The chromosome complement of a normal male, 46 XY.

The double-Y (47,XYY) male is the most interesting of all the chromosome anomalies to the general public. His is the most frequent of the chromosome aberrations — 1 in 250 newborn males, or about 850,000 cases in the U.S. [Stine, 1977, p 153]. Like the 47,XXX anomaly, he is usually fertile and generally indistinguishable in the general population. The public is interested in the double-Y male because of the association between this karyotype and some kind of mental difference that results in a somewhat higher rate of commitment to correctional institutions. The double-Y male tends to be taller than the average male, but this characteristic does not seem to contribute to the social defect. Christensen and Nielsen [1973] studied 10 institutionalized XYY males and found their IQ values to be normally distributed, while the pattern of psychological deviations did not resemble that of any other known anomaly. Their difficulties in social adjustment may start in childhood with various kinds of irresponsible behavior.

The double-Y condition brings up an important legal problem as to whether such a person is responsible for his behavior. If the double-Y male has murdered

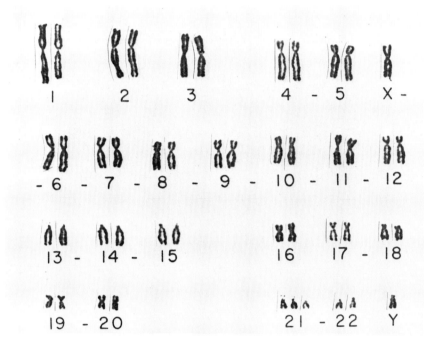

Fig. 3. The chromosome complement of a Down syndrome male, 47, XY, + 21.

someone, should society be protected by sending him to a mental hospital because of his double-Y, or to prison? The decisions will be made on a case by case
basis, but hopefully the public will learn that mental hospitals are more likely to
help the ailing person than are prisons.

The double-Y male is of interest to the genetic counselor because if a chromosome screening of a newborn population is carried out, the XYY anomaly is the
one most likely to be found. The ethical question arises at once as to whether the
unsuspecting parents of the baby should be notified of the finding. Would such
knowledge be unnecessarily traumatic for the parents and cause them to treat the
child as a defective, whereas it might otherwise develop normally? An understanding of the chromosomal aberration has been helpful in providing suitable medical
management of the Turner and Klinefelter syndromes, so probably it will be
possible to help with the double-Y, once the hormonal deviancy, if one exists, has
been studied properly. My recommendation would be that the parents should be
informed of any major chromosomal aberration. There are many reasons for this

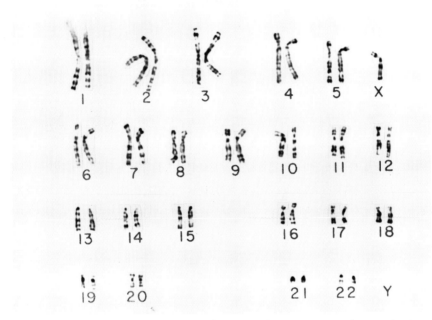

Fig. 4. The chromosome complement of a person with the Turner syndrome, 45,XO.

recommendation and one of them is that, if the affected child develops into a fertile adult, there will have to be a decision made by him as to reproduction. The theoretical risk of the affected person having chromosomal anomalies among his offspring is 50%, though the still undetermined actual risk would be lower. However, the risk is high enough that the person with a chromosomal aberration has a strong ethical right to that knowledge. Borgaonkar and Shah [1974] have provided a thoughtful review of the large literature on the XYY chromosome male, which should be read by anyone involved with a person affected with the condition. See also, Witkin et al [1976] and Hunter [1977] for a discussion of the behavioral aspects of the XYY male.

Dorus et al [1976] screened a sample of 471 enlisted men, 183 cm or taller, serving in the U.S. armed forces and found two 47, XYY males or 1 in 236 of these males. There are more XYY males in a population of tall noninstitutionalized males than is the case for all males at birth. They did not find any striking personality differences in the two males compared with normal values.

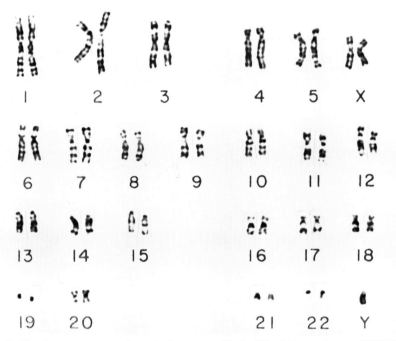

Fig. 5. The chromosome complement of a person with the Klinefelter syndrome, 47,XXY.

The parents of XYY sons are the ones whose opinions (as to whether informa-tion should be given to them or withheld) must be canvassed. This has not been done on an adequate sample, but a letter from a mother, Alice W. Franzke [1975], is very explicit about the matter. Her son, 21 years of age, is certainly not a criminal. She says. "He is a gentle, nice guy, somewhat inadequate occupationally and socially, but always doing his best to succeed. Since birth he has been develop-mentally atypical, and his behavior has been impulsive. He had behavioral, speech, and learning problems in childhood. We were not aware of his genetic make-up until he was 16 years old. . . . If we had known at an early age that he had the XYY genetic make-up, we could have provided more adequate help for him instead of traveling through mazes of misdiagnoses and mistreatment for 16 years. The emotional damage of 'not knowing' has been enormous for the child and his family. . . . Early diagnosis appears to offer more benefit to these boys and their families than does the lack of a possible stigmatizing label."

The letter of this mother contains other comments on telling parents about XYY sons, but the main thrust is what has been my experience in 30 years of genetic counseling, which is that it is best to tell the truth and that the truth does not hurt as much if told in a sympathetic and compassionate way. It takes a lot out of the counselor, as it is stressful to him to convey bad news, but it is of benefit to the parents and the child in the long run. As a general rule, nothing good comes from concealing information.

For a more detailed description of the subject of sex chromosome aneuploidy, the reader is referred to the book by Robinson et al [1979] sponsored by The National Foundation — March of Dimes. The summary indicates that about 2.1 newborns out of every 1,000 have a sex chromosome anomaly. Consequently, there should be about eight million people in the world with a sex chromosome aneuploidy. This vast number of persons with a gross sex-chromosome defect is greater than the total population of many individual states and countries. It is to be hoped that every physician who examines children will use the book in order to help obtain a feeling for the diagnosis of the child with a chromosome anomaly in order that a chromosome study will be initiated for those affected.

Autosomal Chromosome Anomalies (Aneuploidy)

There is, of course, one pair of sex chromosomes in each normal cell of a person and 22 pairs of autosomal chromosomes. It is therefore of interest that over half of the viable chromosomal anomalies are those of the sex chromosomes, and they total more affected persons than all those with chromosomal defects for the other 22 pairs. Part of the reason for this striking phenomenon is that the individuals with sex chromosome aberrations are usually viable, not only at birth but as adults. This is primarily because we have a dosage compensation effect, known as the Mary Lyon effect, such that only one of the two X chromosomes in females is active in the physiology of each cell as is the case, of course, for males who have only one X chromosome in each cell. In normal females the second X chromosome becomes inactive and lodges on the nuclear membrane where it is identified as the Barr body. This process allows greater latitude in the number of sex chromosomes possible in the cells (with consequent viability of the person) than is the case for any of the other pairs of chromosomes.

It is thought about 10% of all fertilized eggs have a chromosome abnormality, but less than 1% of newborns are so affected. Many of the chromosomal anomalies are lost as spontaneous abortions and presumably are the cause of over half of all of such abortions. Kajii et al [1973] looked for trisomies in abortion material and found at least one case for each of 17 of the 22 autosomes. They did not find trisomies for chromosome numbers 1, 5, 12, 17, and 19. No doubt these latter chromosomes would have been found as trisomics also, if the sample of abortuses had been large enough.

A most valuable and recent study is that of Hassold et al [1978] of the cytogenetic causes of the spontaneous abortions at the Kapiolani Hospital (Honolulu) between April, 1976, and the end of March, 1977. There were 234 successful cultures from the abortuses, and almost one half, 109 (46.6%), were chromosomally abnormal and almost all of the 109 involved numerical aberrations. Trisomies for 15 of the 22 autosomes were seen. As was the case with the Kajii study above, no trisomies for chromosomes 1, 12, 17, and 19 were found. Double trisomies were identified in three abortuses. This is considerably less than the theoretical expectation and probably reflects their decreased viability. The majority of the abortuses were expelled between the 11th and 14th weeks, and fewer than 10% were retained longer than 20 weeks.

The picture is different if the survey is of newborn infants rather than of abortuses. The usual trisomies in *newborn* children are for chromosome numbers 13, 18, 21, and rarely 22, plus the sex chromosome types. A cytogenetic survey of 14,069 newborn infants by Hamerton et al [1975] obtained karyotypes on 13,939 babies; 64 (0.46%) had a major chromosome abnormality and 230 (1.65%) had a marker chromosome. There was one 13-trisomic, three had the 18-trisomy, and 14 babies had 21-trisomy. The 21-trisomies were clearly the most numerous type of trisomy; these are the Down syndrome (mongolism) anomaly. The mother of one of the 18-trisomics had a balanced translocation (18;21). There were, in addition, 24 "normal" infants who had a balanced chromosome rearrangement of some sort, and these 24 are of great interest to the genetic counselor because they are at risk of producing trisomic offspring without being aware of their peril. For a short review of this subject, see Evans [1977].

Translocations

This book is not a text on human genetics and no attempt is made to provide detailed descriptions of complicated genetic mechanisms. However, balanced translocations are a significant part of genetic medicine because they are in a sense mysterious, complicated, and a source of great danger to a substantial percentage of the descendants of the carrier of a translocation.

A translocation, in simplest terms, results from the abnormal joining and fusion of two chromosomes from different pairs, say one from pair 18 and the other from pair 21. Small bits from each of the joining or translocating chromosomes may be lost, but as long as the loss is small, the person in which the attached 18;21 chromosome is present will appear to be normal and unaware that he or she is a translocation carrier. The person is in danger only if reproduction occurs. The two fused or joined chromosomes divide without unjoining and remain joined or translocated as long as they continue to be transmitted from generation to generation. The balanced carrier is normal phenotypically and has nearly the total amount of gene material, but as two of the chromosomes are attached end to end, they count as only one chromosome. Thus the balanced carrier has only 45 chromosomes as seen in the microscope, rather than the normal 46.

Figure 6 shows the karyotype of a female who appears to be normal but has a balanced translocation between chromosomes 13 and 14, or 45,XX,t(13;14).

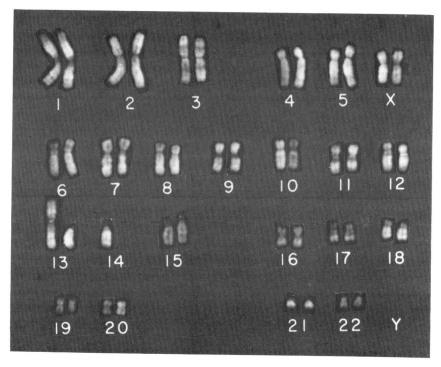

Fig. 6. The karyotype of a female with a balanced translocation for chromosomes 13 and 14, or 45,XX, t(13;14).

She has a balanced translocation. The total count would be 45 chromosomes instead of the normal 46, because the joined 13 and 14 count as one chromosome.

The transmission of the translocated chromosome to the next generation brings up an important ethical question. Some of the offspring receive the translocation in the balanced condition and have the same risk of producing trisomic offspring as did their parent. Should this risk be inflicted upon them and upon subsequent generations, or should the balanced carrier refrain from having children? That is the ethical question. But what is the risk of producing a trisomy if one is a balanced carrier and, therefore, phenotypically normal? The gametes formed by the balanced translocation carrier may contain one normal 21st chromosome as well as a translocated 21st chromosome, for example, and at fertilization another 21st chromosome will be introduced, giving a total of three members of the 21st chromosome, and the "unbalanced" clinical Down syndrome will result. This Down syndrome child resulting from a translocated 21st chromosome will have

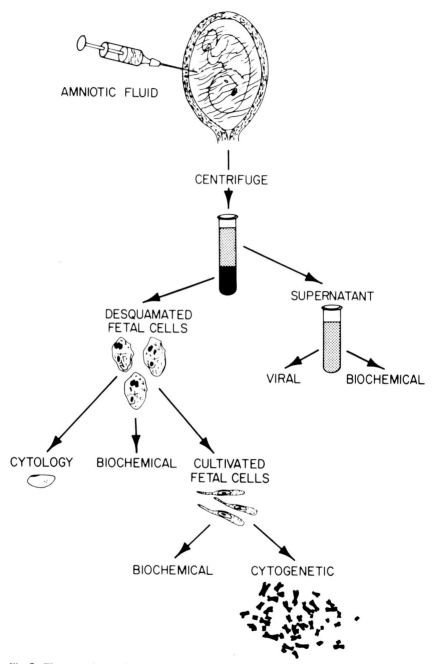

Fig. 7. The general procedure of amniocentesis. The aspiration of amniotic fluid containing fetal cells that are suitable for chromosomal, histochemical, and biochemical studies for the diagnosis of genetic diseases (after Nadler [1969]).

only 206 cases were at high risk for inherited metabolic disorders. However, this latter small group yielded by far the highest percentage of affected fetuses (25%). A total of 297 fetuses (4.8%) in the whole group were affected as shown by the prenatal tests, and in all but five instances the parents asked for interruption of the pregnancy. This last figure indicates the intense desire of the parents-to-be that every child they have will be born normal, if that is possible. The large majority of the 6,121 cases of the parents-to-be, who got the good news that the monitoring showed their fetus to be normal, certainly would consider the modest effort and expense involved to be completely justified.

Another feature of the Galjaard [1976] review is that repeated punctures were necessary in only 4.3% of the cases, and failures of all cultures were reported in only 3.5% of the cases. Galjaard wondered why certain groups at low risk ask for early amniocentesis and why so many others at higher risk do not come for prenatal monitoring. Presumably the answer is simply that education and information organizations are inadequate.

Many centers are able to grow the cells obtained from the amniotic fluid and study the fetal chromosomes. Naturally maternal cells must be avoided, and none of the steps in the process are to be carried out by amateurs. However, the chromosome work is simpler than the monitoring for inborn errors of metabolism. About 75 diseases can now be monitored in a practical fashion but great expertise is required; there are, for example, at least six distinct types of mucopolysaccharidoses (Hurler, Hunter, Sanfilippo, Morquio, Maroteaux-Lamy, and Sly) which are monitored. It is important that you know where the nearest center is where such work can be done. Hopefully, there will be some sort of organization of centers so that only a few will specialize in each disease. The stakes are too high to countenance inferior performance in prenatal diagnosis!

Let us consider briefly a few details about trisomy 21, the Down syndrome, and the risk of the "accidental" nondisjunction, which is the usual cause of it. When the extra third chromosome results from accidental causes we would not expect the same accident to happen twice in the same person. This is the answer, then, to the most frequent genetic counseling question. A couple with a trisomic child (an extra chromosome) should expect to have another only as a coincidence, and the chance of this would be less than one percent. There continue to be suggestions in the literature that the risk is greater than that which is due to coincidence, see Bell and Cripps [1974], but the risk is still small and certainly less than 5% — a risk that most parents are willing to consider favorably, even though it is not insignificant.

Hook and Lindsjö [1978] have provided beautiful curves for the risks of Down syndrome by single-year, maternal-age intervals, which show a sharp rise in the risk from about age 33 to a risk of about 8%–9% in the late 40s. Fortunately, few women have babies when in their late 40's. Hook and Chambers [1977] use such risk figures in calculating cost-benefit analyses of prenatal diagnosis programs for Down syndrome. One of their results is that it is economically feasible to screen

Philosophy

One could ask whether the arguments that support the early abortion of genetically diseased fetuses also support infanticide of the same infants who slipped through the amniocentesis screen and have been born? It is not the same question. These are two different questions. The first question concerns the early fetus which is totally dependent upon the mother and cannot survive elsewhere. The second question concerns a baby which can be raised by the mother or someone else and, therefore, must be accepted as a rightful member of society. It has not committed any crime against society, which would call for its imprisonment or execution. Development of the fetus is a gradual process but cutting of the umbilical cord signals a new status for it. Birth is a good place to draw a moral line; usually it is difficult to draw the line on any continuous variable, but birth as as determined by cutting of the cord is a definite point on the developmental curve.

A frequent question is when does life begin? Life does *not* begin with the fertilization of the egg. It began millions of years ago and has continued ever since then. It never starts anew. Both the egg and the sperm are living material, as is obvious to anyone who has seen them through the microscope. How long each person's life lasts is entirely conditional upon the genotype of the person and the environment it experiences. Death can occur at any time and is the usual fate of all unfertilized eggs and sperms. All life is an extension of preexisting and continuing life. The egg has the potentiality of *becoming* a human being after elaborate developmental changes have occurred. But as a simple human cell it has none of the attributes of a human *being* and does not deserve to be considered a human being. The question becomes, "At what point does this blob of cells *become* a human being?" The answer is a purely arbitrary determination, which results from one's philosophy.

The genetic counselor has an indirect interest in the subsequent mental health of the mother who has chosen an abortion for genetic reasons. Greet et al [1976] followed for two years a series of 360 women who had first trimester vacuum aspiration terminations. These were not abortions primarily for genetic reasons, but the study is the most appropriate one known to me. Each woman had received brief counseling before termination. The detailed, structured interviews at 3 and 18 months were of good quality. Compared with the ratings of psychosocial adjustment before termination, significant improvement had occurred at follow-up in respect to psychiatric symptoms, guilt feelings, and interpersonal and sexual adjustment; there was no significant change in marital adjustment. Adverse psychiatric and social sequelae were rare. Many reports related to abortion are of an emotional nature, but this one seems to be free of bias. One might make the comment that the psychologic state of the women would have improved if the fetuses had come to term and the babies were born. The point is that the termination did not damage the psyches of the mothers; rather, there was improvement − for whatever reason.

The usual result of prenatal diagnosis is that the fetus is found to be unaffected and the relieved parents happily await the birth of the baby. Before prenatal diagnosis was developed, worried couples sometimes aborted the fetus without knowing whether it was affected or normal. Such needless abortions of normal fetuses can be prevented by using the amniocentesis technique. However, if an abortion is decided upon as a result of prenatal diagnostic data, the abortus must always be examined to confirm the prenatal findings. This is important both for monitoring the accuracy of the prenatal tests and for assuring the parents that the results of the test were correct. Riccardi [1977] pointed out that, "A clinician would be remiss — professionally, ethically, morally and legally — to deny a couple the opportunity to obtain information about the fetus or to implement a decision about continuing the pregnancy. Moral decisions about abortion must be made by the couple involved, not by the clinicians assisting them."

7
Treatment for Some Genetic Diseases

The treatment used to ameliorate genetic diseases is not a part of genetic counseling or even of the broad topic of genetics. Genetics is the science of the transmission of the genes from parents to offspring and not that of subsequent efforts to overrule the effects of the genes by some type of therapy. However, the genetic counselor includes numerous services in his repertory and has an implied responsibility to refer his counselees to the appropriate clinic or agency for help. Usually, the process works in the reverse direction: the physician refers the patient being treated, or the parents of the patient, to the genetic counseling center. Thus the diagnosis has been established and treatment initiated in many cases perinatally, and the genetic counseling comes later when a subsequent pregnancy is contemplated.

The genetics of several hundred inborn errors is known, but treatment is available for relatively few. The preventing of the conception of children with inborn errors of metabolism is the most desirable way to avoid the inconvenience and difficulties of the treatment. For most traits no treatment is known, so prevention of the conception of the abnormal child is clearly the easiest goal to pursue. However, most heterozygous persons are not aware that they are carriers for any particular recessive gene, and they are astonished when a child with an inborn error is born to them. They insist that no previous example of the trait has occurred in their ancestries, which is no surprise to the geneticist. Unless some screening system for all carriers in the population for a specific trait is developed, we will have the majority of the cases with a recessive trait appearing in families as the first affected member of the kinship within memory.

Phenylketonuria

The discovery that the mental retardation associated with untreated phenylketonuria can be prevented by rearing the affected infant on a diet low in phenylalanine opened up the area of treatment for the inborn errors of metabolism. Though the mean IQ of early treated patients tends to lie about 10 points below that of the normal population, they are educable at regular schools and differ strikingly from most untreated phenylketonuric children whose mean IQ is about 50. It is thought that high phenylalanine early in life can cause marked changes in the composition of the brain and that it inhibits the transport and concentration of other amino acids in liver and brain. Actually the cause of the mental retardation is far from understood.

The dietary treatment of a metabolic disorder such as PKU is highly complex and potentially hazardous. Presumably no physician would embark on the care of such patients unless he or she is able to work closely with a dietician experienced in the treatment of the trait. For the best results the babies should be treated in centers designed for the purpose. The length of each admission should be kept short and there should be good cooperation among all concerned.

There are two religions in the United States as to how long the patient should be kept on the low phenylalanine diet. One states that females should be kept on the diet until menopause, but the other realizes that some more realistic stopping point must be accepted. Once the child is off the diet and has experienced the joys of ad lib eating, it is difficult for him to get back on the diet.

The children born to mothers with untreated phenylketonuria have a very high incidence of congenital abnormalities, including mental retardation. These children are obligate carriers of the gene for PKU, and ordinarily would not be affected. However, the high phenylalanine levels of the mother, to which they are subjected in utero, cause irreversible damage to their brains. The only solution for homozygous mothers with high phenylalanine levels is planned pregnancy with dietary restriction throughout the pregnancy. The diet during pregnancy presents many problems and is potentially hazardous. The reader is referred to the book edited by Raine [1975] for a more detailed treatment of these problems.

Homocystinuria

There are two main approaches to treatment for Mendelian recessive gene defects. The first is the restriction of substrate accompanied by product replacement as in PKU, and the second is by co-enzyme supplementation. In the latter case the defect may be in the vitamin precursor of the co-enzyme. Homocystinuria is an example of a trait which can be influenced by supplementation with vitamin B_6. Pyridoxal phosphate is the most important active form of the vitamin, which as co-enzyme takes part in a variety of enzymatic reactions involving amino acids. There is no evidence of any deleterious effect of high doses of pyridoxine given over prolonged periods, either to children or during pregnancy. Therefore, upon diagnosis of a patient with homocystinuria, a trial of pyridoxine should be given irrespective of age. It should be known within three weeks if the patient is responding to the treatment. We cannot go into the details of testing and treatment here, and again refer the reader to the book edited by Raine [1975].

Homocystinuria is a very rare disease, much less frequent than PKU. However, it is a slowly progressive disease and treatment is effective to some degree in older children. It is very interesting from a genetic counseling point of view as prenatal detection is possible. Cystathionine synthase is the enzyme involved, and its activity level can be determined in both skin cells and in the fibroblast cell lines derived from cells in the amniotic fluid. Screening of a population could be done to detect the heterozygous persons, but the cost-benefit ratio probably is not favorable enough to make this worthwhile, except for relatives of an affected person. In case there has been an affected child, the parents should be interested in amniocentesis if a pregnancy occurs and the siblings and aunts and uncles of the affected person should be tested for heterozygosity for the gene. Elective termination of the pregnancy is feasible if the fetus is found to be affected. One should beware of confusing the symptoms of this disease with those of the Marfan syndrome.

Maple Syrup Urine Disease

Maple syrup urine disease should be mentioned briefly because it is an emergency situation. The rapidity with which irreversible damage to the central nervous system can occur makes early diagnosis and treatment imperative. It is frequently diagnosed too late when it is a first occurrence in a family. Often valuable time is lost in obtaining such procedures as EEG's and PEG's instead of pursuing possible metabolic causes for the central nervous system symptoms. The odor of these patients and their urine is quite remarkable and provides the label for the syndrome. The 'nose test' should be part of the screening of all urine. More precise tests are available than that of the maple syrup smell, but speed is of the essence in getting the baby onto a low protein specific diet. Treatment is based on a limitation of intake of the three branched chain amino acids involved. Top scientific assistance is needed and, if initiated early enough, there will be good results and death will be prevented. Careful monitoring during treatment is necessary. It is not known how long the dietary restrictions must be maintained.

There is a vexing ethical question involved with maple syrup urine disease. Should one institute treatment if the diagnosis is not made until after severe neurological damage has already occurred? The damage is irreversible and without treatment the child will die, usually within a few weeks. Are the parents legally free to decline treatment for the severely damaged child?

Wilson's Disease

Wilson's disease was one of the first inborn errors of metabolism for which treatment was found. Many patients undergo a long series of investigations before the correct and very clear diagnosis is made. The major reason for the delay in diagnosis is that it is such a rare disease and, of course, the symptoms are ill defined at first. However, any child with jaundice for which there is no obvious cause should be considered a possible case of Wilson's disease, and the concentration of copper and ceruloplasmin in the serum, the urine, and perhaps in the liver must be determined. The sooner the diagnosis is made, the more successful the treatment. The treatment consists of the administration of D-penicillamine, which will mobilize large quantities of copper for excretion from the body. There is no standard dosage and there is, of course, no permanent cure. The defect is genetic and therefore lifelong, no matter how helpful the treatment is. The results of treatment are usually very rewarding, though there are occasional failures and the patient dies. The reader is again referred to the book edited by Raine [1975] for the mass of details relating to the treatment of this recessive trait.

Fabry's Disease

The list of inborn errors of metabolism that can be treated would be of great interest to Sir Archibald Garrod, who was the first to comprehend the situation. We cannot consider all the diseases on this list, but should at least mention Fabry's disease, because it is distinctive from those already considered and is of especial interest here at the Dight Institute where it was a major research objective of Dr. Robert Desnick and his associates.

Fabry's disease, angiokeratoma, is of considerable genetic interest because it is X-linked. The gene shows measurable linkage with the Xg^a blood antigen gene. The symptoms of the disease result from the progressive accumulation of the glycolipid, galactosylgalactosylglucosyl ceramide, in most tissues of the body. The metabolic abnormality results from the absence of activity of α-galactosidase. The enzyme is required for the catabolism of the trihexosyl ceramide. The absence of activity of the enzyme in the hemizygous males results in the deposition of the lipid in various tissues and particularly in the blood vessels of the heart and the kidney. This causes crises of extreme pain in the extremities and ultimately leads to heart or kidney dysfunction. Death usually occurs in adult life from renal failure or cardiac complications. Heterozygous females may exhibit the disease in an attenuated form and are likely to show the characteristic corneal opacities. Telangiectases may be one of the earliest symptoms and may lead to diagnosis in childhood. Treatment should start at once in order to prevent damage from the stored material.

Attempts have been made to replace deficient α-galactosidase activity by infusion of normal plasma. Measurable levels of enzymatic activity were found in the plasma of these patients, and the activity rose in six hours to 150% of the average found in normal plasma. Loss with rapid turnover during the first day resulted in no further activity by the end of a week. Recent efforts have included intravenous injection of highly purified α-galactosidase; these infusions have been extremely encouraging and it is proposed that weekly infusions will be therapeutic for a lifetime disease. Attempts were made to transplant kidneys, which would function as manufacturing plants to produce ceramide trihexosidase. This was successful, but involves host acceptance of the kidney and is a rather heroic procedure. Sophisticated but hopefully easier ways of delivering the enzyme are being researched at present.

Many Successes

Limited but significant success has been achieved in increasing substantially the life expectation of the child with cystic fibrosis, as will be described in Chapter 13 of this book. Transfusions for the hemophilic male increase life expectation, surgery for neural tube defects is often successful and almost always succeeds with pyloric stenosis. There have been splendid advances in the management of seizures and diabetes mellitus. Administration of large amounts of substrate may bring about "normality" for the patient with vitamin-D-dependent rickets if provided early enough. The prevention of death from rhesus factor incompatibility is a well-known example of the important discoveries that have led to the outstanding victories in the treatment of genetic diseases.

Genetic counseling is not the treatment of genetic disorders, but the counselor can be instrumental in guiding people to the expert in the treatment of the particular genetic disease of concern to them.

8
A Few Laws

There is no cause for alarm! The few laws to be mentioned are simple in principle and the arithmetic will be simplicity itself. The physician does not need to be a statistical genius to understand the necessary principles of medical genetics as they apply to genetic counseling. Human genetics includes medical genetics but goes far beyond the latter in its statistical requirements. Much of the material published in the *Annals of Human Genetics,* for instance, has mathematical content that cannot be read except by a few specialists in human genetics. It serves its purpose but is usually too specialized to be useful to the family physician. This book is written primarily for the family physician and, hopefully, is restricted to material relevant to his needs.

The Powerful Number 2

The laws of heredity are statements of the random assortment and recombination of the pairs of genes present in the ovary and the testis and then joined in the fertilized egg. As the chromosomes, and the genes within them, occur in pairs, the laws of heredity are based upon various manipulations of the number 2.

Let me give a personal example related to this powerful number 2. One of my ancestors, William Reed, came to this country in 1635 with his two sons William Jr. and George. If each of them had two offspring who reproduced, and so on down through the generations to my own generation (which is the fifteenth), then there would be 32,768 descendants of the original William and his wife. This seems reasonable, but if we assume that he had four children who reproduced and each of them had four reproducing children and so on for fifteen generations, then there would be 1,073,741,824 descendants of William Reed as of today. This figure is utterly absurd, as the whole population of the U.S. numbers only about one quarter of this total. These strikingly different totals, resulting from whether there was an average number of two reproducing offspring or four of them, tell us a great deal about medical genetics. They emphasize that it is the reproducing offspring who relate to subsequent generations and not those persons who fail to have children. Many of the medical genetic traits are at a disadvantage and are at low frequencies because of that. Other traits, like the blood groups, seem to be advantageous under some conditions and disadvantageous in other situations, thus maintaining rather high frequencies over the generations. These polymorphisms are "normal" traits rather than medical conditions in many cases.

Autosomal Dominant Inheritance

If the statistics are collected for all the children of a group of patients with Huntington's disease (chorea), it will be found that about half of these children will also develop the disease when they reach the age for its appearance. No further cases will appear in the descendants of the normal children, but if the affected persons produce children about half of them will be stricken with the disease eventually. In each succeeding generation normal individuals will always produce normal offspring, but one-half of the offspring of the affected individuals will develop the disease. If each of our patients with Huntington's disease had 2 children on the average, one who became affected and one who did not, there would be only one patient with the disease to 32,767 without the gene for it after 15 generations. This assumes that there is no reproductive advantage or disadvantage due to the presence of the Huntington's disease gene, a difficult assumption to prove. However, we do know that the choreic passes the gene down through the generations in many cases. The affected descendants get the chorea gene from the affected parent and a normal gene from the other parent. Because the chorea gene "dominates" the normal gene and causes the disease to appear, it is called a *dominant* gene or trait. The normal gene partner which is suppressed or concealed by the chorea gene is appropriately named a *recessive* gene.

It is hard to think of normal genes being in the recessive state and a dominant gene being the case of a disorder. Our mind set is that almost all of our genes are dominant genes for good health, and that genes for disorders are carried by a few persons — and when two of them carry the same gene they have a chance of producing an affected child. This is often the case but the list of traits in which the disorder behaves as an autosomal (not on the X-chromosome) dominant is very long. McKusick [1978] lists 1,489 autosomal dominant traits. In all of these cases, the gene for normal development and good health is recessive to the dominant gene for the anomaly. He also lists 1,117 autosomal recessive traits and 205 X-linked traits. All three mechanisms total 2,811 traits.

It should be remembered that because the gene for an anomaly or disorder is at a disadvantage in most environments, it will be at a low frequency in the whole population. Consequently, few people will possess the deleterious dominant gene on *both* members of the relevant chromosome pair. Almost always the affected person will carry the normal recessive allele or gene partner as well as the dominant gene which is responsible for the patient's anomaly. Thus only half of the children of an affected person are affected likewise. Figure 8 illustrates the transmission from generation to generation of a gene responsible for an autosomal dominant trait.

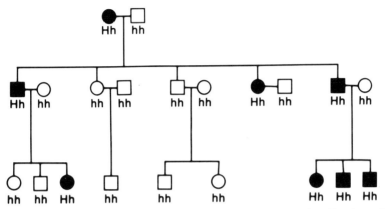

Fig. 8. A pedigree demonstrating the inheritance of a dominant trait such as Huntington's disease. The circles represent females, the squares males. Filled in symbols indicate that the person is affected with the disease and as it behaves as a dominant trait only one gene (H) for the trait is necessary in order that it be expressed. Note that there is transmission from affected father to affected son which eliminates the possibility of X-linkage.

Autosomal Recessive Inheritance

The gene for a disease is not always dominant to the normal member of its own gene pair. The gene for albinism is recessive to its normal gene partner. An albino has the defective type gene on *both* members of the relevant gene pair, so the trait shows itself. If an albino produces children with a normal person as the other parent, we expect the children to have normal pigmentation. Each of the children must carry the albino gene that was received from the albino parent, that is, the child is an obligate carrier. The carrier child looks normal because its recessive albino gene has been compensated for by its dominant normal partner gene.

When both parents carry the gene for albinism, simple random combination of their normal and albino genes gives a predictable ratio of albino and normally pigmented children. If the dominant gene for normal pigmentation be indicated by C and the gene for albinism be shown as c, the person who has both genes, or Cc, is a carrier of the trait. The sperm produced by this carrier will have only one member of the gene pair in each sperm. Half the sperm will have C and the other half c. The same will be true for the eggs of the carrier woman. When two carriers marry, the ratio of all possible combinations of the C and c eggs and sperm will be one CC child, two Cc carrier children and one cc albino child. Because of the dominance of the normal gene, the CC child, without any gene for albinism, cannot be distinguished clinically or phenotypically from the two carrier Cc children, who also look normal. The ratio we see is three pigmented children to one albino. See Figure 9.

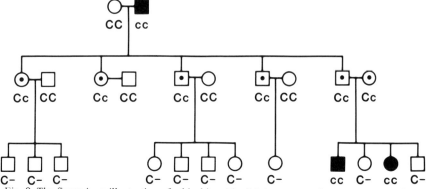

Fig. 9. The figure is an illustration of a kinship with albinism, a recessive trait. All the off-spring of the affected male (cc) are obligate carriers of the gene for albinism as indicated by the dot in the center of the symbol. One of these carriers happened to marry another carrier and produced an albino son and an albino daughter in the next generation. Some of the members of this last generation will be carriers but we cannot tell which ones they are by looking at them. The dash after the symbol indicates the uncertainty as to whether the person is a carrier.

Our albinism model demonstrates the 1:1 ratio as well as the 3:1 ratio above. The 1:1 ratio would be shown in the children from a carrier, Cc, and an albino, cc. Half of these children will be carriers, Cc, and the other half albinos, cc. Half of these children will be carriers, Cc, and the other half albinos, cc.

It should be emphasized that most families are too small to give good genetic ratios. Obviously, a single two-child family cannot by itself give a 3:1 ratio. However, the expectation for *each child* is always precisely that indicated by the appropriate genetic ratio. For instance, if a normal couple have had an albino child they will know that they are both carriers of the gene for albinism and that each subsequent child has a one out of four chance (25%) of being an albino. The fact that they have already had an albino child tells them they are carriers of the gene, but it is no insurance against having additional albino children. Each conception is an entirely independent event as far as the genes are concerned, and is not influenced by previous gene combinations.

The 3:1 and 1:1 ratios, or 25% and 50% expectations are what the family physician should have in mind.

The physician engaged in clinical research needs to have additional information as to the statistical problem, or bias, introduced in his data by the small sizes of the individual families. The bias can be overcome by pooling the data from many families. If an albino child is the first child in the family, then all such one-child families will be composed of albinos only, that is, 100% affected instead of the 25% one would expect when both parents are carriers, Cc. This discrepancy results because there will be three families where both parents were carriers, but the child was Cc or Cc and appeared normal and thus did not come to the clinician's attention.

If we consider the family size of two children, there will be nine families with two normal children which will fail to be detected, six families of one albino and one normal child, and one family with two albino children in it. Thus, only seven out of every sixteen two-child families will be detected, and 57.1% of the children in the seven families will be albinos. Thus there is still a large discrepancy, or bias, from the 25% expectation for the pooled data. However, if by some chance two carrier parents produced as many as 16 children, it is almost certain that at least one child will be an albino. Everyone of these mammoth families should appear at the clinic, and a group of these families would have the average of 25% albinos expected from carrier parents.

The clinician who is doing research on a disease that has not been explored genetically must arrange his data according to family size in order to detect the type of heredity involved. A number of cases are known in which the scientist was so badly confused by the bias due to small family size that the hypothesis of simple recessive genetics was rejected! There are several methods for correcting for the small family size bias. The Macklin method showing the "Percentage Affected Expected" is probably the most convenient for the clinician. He should arrange his families according to size, and each family will have at least one affected child in it. Then find the percentage of affected children for each family size and check to see whether there is reasonable agreement with Table 8-1, given below. At this point it would be well to bring the data to the attention of a professional geneticist who might suggest additional analysis of the material.

The bottom line of the table is appropriate for either a carrier of a recessive trait times an affect parent, or a carrier of a dominant trait times a normal person.

Remember that genetics is not as simple as these neat ratios might lead one to believe. Sometimes when two albinos marry, the resulting children may all be normally pigmented. This would be a classic example of genetic heterogeneity which has been explained in Chapter 4.

X-Linked Inheritance

The third type of regular Mendelian transmission of genes is called X-linked. The sex of the baby depends upon whether it receives an X chromosome from the father or a Y chromosome. The X and Y are easily distinguishable as the reader will recall from the photographs in Chapter 5. Boys receive no X chromosome

TABLE 8-1. The Percentages of Affected Offspring Expected in Different-Sized Families That Come to Their Physician or Clinic

| Parents | Total number of children in the family | | | | | | | | | |
	1	2	3	4	5	6	7	8	9	10
Cc x Cc (%)	100.0	57.1	43.2	36.5	32.8	30.4	28.8	27.7	27.0	26.5
Cc x cc	100.0	66.0	57.0	53.4	51.5	50.8	50.4	50.0	50.0	50.0

from their father, only their Y chromosome. Thus no gene on an X chromosome can be transmitted from father to son. Consequently, the rule is that a gene transmitted from father to son should reside on an autosome, or extremely rarely, on the Y chromosome. Thus, for a trait such as red-green color blindness, we get "criss-cross" inheritance. That is, the color-blind father transmits his X chromosome with the resident color-blind gene to all of his daughters but to none of his sons. The carrier daughters transmit the X chromosome with its gene for color-blindness to half of their sons who will be color-blind. The other half of their sons will have a normal X chromosome, which came from their maternal grandmother.

It is difficult to understand how the three Mendelian mechanisms work without a little practice. It is hoped that the figures, such as Figure 10 for X-linkage, will be helpful.

It should be obvious that it will be easier to discover the major type of inheritance involved (such as dominant, recessive, or X-linked) if the trait is congenital. A major challenge to the physician is that of detecting phenlyketonuria, cystic fibrosis, and so on as close to birth as possible in order that therapy or planning of some sort may be instituted. If the trait does not express itself during the early months of life, we have another bias to contend with due to the variable age of onset.

Let us consider a person whose parent suffered from a dominant type of ataxia. The a priori probability that the person inherited the gene for the ataxia is, of course, 0.5. We know this to be true because the person had one affected parent. The person is now 50 years old and has not shown any signs of ataxia, whereas

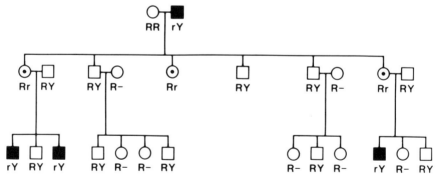

Fig. 10. X-linkage for a recessive trait is of the criss-cross type. Let us start with a male affected with red-green color-blindness. We use the symbol (rY) indicating that the gene for color-blindness (r) is on the X-chromosome and that there is no corresponding gene on the Y chromosome. All of the daughters of the affected male will carry the gene (r) but none of the sons will have it. Sons of affected males do not get the gene (r) from their father so they do not have or transmit the trait to their descendants. The dash after the symbol indicates the uncertainty as to whether the female is a carrier.

the average person with the gene will have shown symptoms by about age 35. Our counselee has thus safely passed much of his risk period. It could be calculated that, in fact, our person may have only a 5%–10% risk remaining. This should be very pleasant news for the person, especially as he or she may have worried a great deal that the gene for ataxia might have been transmitted to some of the children.

The traits that the physician comes in contact with most frequently are not rare disorders such as ataxia, but frequent ones such as diabetes, which is not detectable at birth and has no simple genetic mechanism as its cause. Such heterogeneous traits also depend upon important environmental factors for their expression. The genetic background may be sufficient for expression of the trait under some environmental constellations but not under other constellations. Sometimes, because of genetic and/or environmental differences, the trait will fail to express itself and is said to *lack penetrance*. Traits also may have different *degrees of expressivity*, an example of which would be cleft lip, which can vary from a slight nick in the lip unilaterally to complete bilateral clefts. This variation would be the result of the interaction between various genetic and environmental constellations.

Autosomal Linkage

One of the principles developed early in the history of genetics is that of linkage. We mean by genetic linkage that many of our gene loci must be linked with each other because they have to be located on the same pair of chromosomes. We have only 23 pairs of chromosomes with thousands of gene loci located upon them. It has been a tremendous job to locate the linear positions of some of the many loci on some of the chromosome pairs. Furthermore, one should keep in mind the distinction between gene linkage (which is measured in terms of crossing over between the partner chromosomes) versus gene or trait associations (such as found between the ABO blood group genes and various cancers.) The gene associations do not imply any physical relationship of the genes within the cell, but instead a physiological interaction. These associations are usually too loose to be helpful in genetic counseling. Most of the genetic linkages are also too loose for genetic counseling purposes. It is easy to predict that eventually a genetic linkage will be found between Huntington's disease and some trait like a blood group that is so tight or close that one can identify the future Huntington's patient at any time after birth or even in utero. This technique will be useful for counseling only if the gene linkage between the gene for Huntington's disease and that for a blood antigen is so close that crossing over almost never occurs. Loose linkages could easily result in misconceptions and subsequent harm, in counseling. The subject is complicated enough that the physician could well take linkage situations to a counseling center for the latest information about specific linkage groups data that changes almost every day.

Genetic counseling has become increasingly complicated since I coined the term in 1947. Even the Mendelian ratios considered in this chapter give probabilities that need modification when the trait has varying ages of onset in different members of a family. These modifications of the basic ratios are treated in elaborate detail in the book by Murphy and Chase [1975]. The major counseling difficulties do not result from modifications of Mendelian ratios, which can be corrected in some straightforward mathematical fashion. They result from the "environmental" factors and modifying genes, which are always of substantial importance when dealing with the traits the family physician is most likely to encounter. The risks relating to these frequent traits, such as spina bifida and so on, will be developed later on in this book.

Perhaps the most important law is also a commandment, "Beware of the Biases!" If one is aware of the biases involved in a problem, they can be reckoned with in some way. The difficulty comes when we don't know that our conclusions are biased. An interesting paper on this subject is that of Ten Kate [1977] regarding the high frequency of heterozygotes for cystic fibrosis, Tay-Sachs disease, and phenylketonuria in specific ethnic groups. He demonstrated very neatly that in such cases there are several probable sources of bias: on the one hand there is the fact that ascertainment of heterozygotes through affected offspring will tend to underestimate the relative frequency of smaller families, and on the other hand there are the inadequacy of census data for comparison and the biases inherent in the selection of control families.

Biases are important in leading the counselor astray but they are not the only source of error. The failure to construct a valid family pedigree may also result in a counseling disaster. This does not mean that one has to unearth all the person's ancestors back through 15 generations, but it does mean that you have to search out information for all the relevant persons concerned with the trait presented for all the relevant persons concerned with the trait presented for counseling. The amount of information needed will depend upon the kind of trait involved, and there is no recipe stating the number of relatives for which information must be obtained. Background experience in human genetics provides the answers as to whether a pedigree must be obtained and how extensive it should be.

It is my hope that by now the reader has developed a healthy respect for the complexities of human genetics and will have no false pride in asking for confirmation of the conclusions reached prior to providing genetic counseling, or even afterward. In many situations there is no clear answer to the counselee's question. To reply, "I don't know," is one of the most honest and conscientious answers in the scientific world.

9
The Ubiquitous Heterozygote
(or, the Common Carrier)

Typhoid Mary was a common carrier, but it was possible to isolate her from contact with foods and thus to remove her as a danger to the community. The carrier of a deleterious gene is ubiquitous. If every carrier of a disadvantageous recessive gene refrained from reproduction, there would be no more children and our species would become extinct like the passenger pigeon. Don't worry, reproduction will not cease, even though almost everyone is the carrier of at least one deleterious recessive gene.

The reason that we get normal children most of the time is that the recessive gene or genes that I carry are different from those that my wife carries. We are both carriers, or *heterozygotes,* but not for the same recessive genes as a rule. If we had both carried the same concealed recessive gene there would be a 25% chance for each of our children to get this recessive gene from both of us. The child with the double dose of the gene, rather than just one recessive, is called a *homozygote* and is said to be *homozygous* for the gene. This child will display the trait or phenotype produced by the gene when it is present on both members of the relevant pair of chromosomes.

It is well known that universities grow because all the students bring some information *to* the university but few of them take much *away* from it. The concept that would be most valuable for you to take away with you from this book is the concept of the ubiquitous heterozygote.

Let's illustrate the concept with a recessive trait like albinism. If a couple discovers that both of them are heterozygous for albinism by producing an albino child, they should be taught that the chance is 1 in 4 for each of their subsequent children of being an albino also.

The frequency of albino births in the whole population is 1 in every 10,000–20,000 births. Thus, any couple picked from the phone book is extremely unlikely to have had an albino child. This randomly selected couple has about 1 chance in 20,000 that their next child will be an albino, quite a different picture from that of the 1 in 4 chance for the couple both of whom are heterozygous.

69

Now the surprise! With 1 in 20,000 of the population being albinos, what proportion of the people carry the gene for albinism in the heterozygous condition? The answer is that about 1 out of every 70 persons carries the gene for albinism! The next time you are in a large crowd of people you might remember that 1 in every 70 of them is heterozygous for albinism and probably doesn't know it. A simple bit of arithmetic will show you that the heterozygote, or carrier of albinism, is 286 times more frequent than the homozygous identifiable albino. Thus, the chance that I carry the gene for albinism is 1 in 70 and that my wife carries this gene is likewise 1 in 70. The chance that we are both heterozygotes is not the sum of these two probabilities or risks, but the product of them, or 1 in 4,900. We multiply the 1 in 4,900 by 1 in 4 which is the Mendelian risk of getting a homozygous child from two heterozygous parents, or 1 chance in 19,600, or roughly 1 in 20,000 – the frequency of albino children in the whole population.

It is easy to see that the more rare the recessive trait, the greater the disproportion between the frequency of carriers and affected persons. Such calculations are made by using the Hardy-Weinberg law. This law is valid for use with large populations that have been genetically stable for a long time, where no consanguinity is present, and in which there has been no assortative mating for the pair of genes under consideration. No population ever meets all of these criteria with precision, but the law provides extremely useful approximations, if applied intelligently in appropriate situations.

The Hardy-Weinberg law states that there is a predictable relationship between the frequency of heterozygotes and the frequencies of the two homozygous classes in the population. If the frequency of the normal gene, C, is indicated by the algebraic symbol, p, and the frequency of the gene for albinism, c, is indicated by the symbol, q, then the binomial expansion of p + q gives the relative frequencies of the three genotypes. Thus, p^2 (CC, homozygous normals): 2 pq (Cc, carriers): q^2 (cc, homozygous albinos).

The genetic diseases that appear at the Dight Institute often enough to justify a chapter to themselves in this book usually have frequencies greater than 1 in 1,000 births. If one recessive is necessary in the homozygous state for the disease to appear in a frequency of 1 in 1,000 births, then at least 1 out of every 16 people would be heterozygous for this gene as can be seen in Table 9–1 below. Time after time, parents protest that the defect in their child could not have a genetic basis because they are sure that nothing like it had been present in their families. They do not realize that the recessive gene involved could have been passed down through their ancestors for 100 generations or more without having become homozygous. Furthermore, no one knows what ailments were present in distant relatives in the past.

The most practical use for human genetics is in genetic counseling. Unfortunately, at present most counseling occurs "after the fact." The couple has already produced a child with an anomaly and wants to know what is the risk of having

TABLE 9-1. The Ratio of Carrier to Affected Individuals for Cases of Single Factor Recessive Inheritance (after Stern, 1960)

Frequency of affected persons in the population	Frequency of carriers in the population	Ratio of carriers to affected persons
1 in 10	1 in 2.3	4.3:1
1 in 100	1 in 5.6	18:1
1 in 1,000	1 in 16	61:1
1 in 10,000	1 in 51	198:1
1 in 100,000	1 in 159	630:1
1 in 1,000,000	1 in 501	1,998:1

another. Ideally, genetic counseling should be premarital so that if both members of a prospective union turn out to be carriers of the same gene they could plan for the future.

But how can the carrier be detected before he has produced an affected off-spring? I often have couples phone me who are planning on getting married, and they ask if they can be tested for whatever disadvantageous genes they might be carrying. The answer is no, as a generalization. There isn't much reason to give test after test unless there is some expectation that a specific gene might be present in the heterozygous condition in both members of the couple. On the other hand, screening programs are anxious to test everyone who fits into their purview. If one or both members of the couple were Jewish, our Tay-Sachs screening program would be glad to test them. The cost-benefit ratio becomes important in screening for carriers of a particular trait because the likelihood of a positive result of the test is so small for any couple picked at random. Thus if the gene frequency for a rare recessive is 1 in 100, in only 1 in 2,500 couples would we expect to find both members to be carriers.

It is appropriate to screen for homozygous recessive traits in a random or entire population in order that the affected persons can be found and treated. In the ideal world one should test for all traits for which techniques are available. However, the cost is so high for most of these special tests at present that we are willing to forego the benefits of screening in most cases.

It should be clear that the risk of both members of a couple carrying a specific recessive gene is very small. The number of recessive genes which one could carry is large; the McKusick [1978] catalog of different recessive genes contains over 1,000 of them. If we guess at an average frequency of 1 in 100 carriers for each one of them, each person then could carry 10 of the 1,000 genes. Thus everyone is heterozygous for 10 genes, on the average, and the heterozygote is ubiquitous indeed.

There is considerable information from blood group genetics and various in-born errors of metabolism as to the frequency of the different alleles composing these polymorphic systems. The different alleles at any one locus are often co-dominant rather than behaving as recessives, but this makes it easier to detect heterozygosity, which is our concern here. Harris [1970] estimated that in any one individual, 6% of the gene loci were heterozygous for the enzymes he studied, while Lewontin [1967] followed up a suggestion of mine and found that in any one person, about 16% of the blood group loci were heterozygous. Truly a significant amount of heterozygosity. See Hedrick and Murray [1978] for recent estimates of heterozygosity.

The homozygous state of these common blood group or enzyme genes is part of "normal" human variation and generally would not be expected to be particularly deleterious or lethal in its effects. The general concept regarding polymorphisms is that particular gene complexes are beneficial in some circumstances and detrimental in others. The frequencies of the different alleles fluctuate in response to "environmental" changes over time. The recessive abnormalities of concern to the pediatrician, such as Tay-Sachs' disease, for instance, are different in that they are rare and there is no conceivable situation in which the homozygous genotype would be advantageous. While it is possible that the heterozygous state of some of these gross abnormalities might be advantageous, in most cases the gene frequencies for most of them are probably maintained by new mutations from the normal to the aberrant allele. The life expectancy of a newly mutated gene of this sort may be for many generations as it may be advantageous in the heterozygous condition or just protected by the normal allele which is present.

You now know; you are the ubiquitous heterozygote, as is practically everyone else.

10
Please Don't Marry a Relative
(Consanguinity, Assortative Mating)

This request that people not marry blood relatives is a polite way of pointing out that there are social and biological disadvantages in consanguineous marriages. The marriage of close relatives incites at least some social opprobrium and can result in genetic disaster for some of the offspring of the related couples. These disadvantages are "relative" to the closeness of the genetic relationship between the conjugal partners. The religious and legal prohibitions against consanguineous unions are, of course, the voice of society. The religious and legal sanctions may be poorly enforced but everyone is aware of their existence. Why then do blood relatives marry?

It is well known that assortative mating is a very strong force. It results partly from propinquity as most people marry someone with whom they are associated in some way. People marry others who are of the same race, religion, education, and physique much of the time. The correlations for each characteristic are not perfect but they are usually quite high. It is not clear to me why people are more comfortable with similar rather than contrasting characteristics, but that is the case more frequently than expected by chance. One's relatives are more similar to the person genetically than are nonrelatives, and they are apt to be lifelong acquaintances. Thus relatives are likely to marry more often than one would expect on a random basis. Thus we have a rough equilibrium between the advantages of consanguineous unions and the disadvantages that would tend to restrict them.

The religious and legal prohibitions against consanguineous marriages are usually for those persons more closely related than second cousins. The genetic risks for the offspring of second cousin unions or more distant relationships are only very slightly greater than those for the offspring of unions of nonrelatives. Thus society has perceived where the safety line is for both social acceptance of consanguineous unions and genetic risks. There has been much confusion in the public mind as to the biological consequences of consanguinity or incest, as the closer relationships are called. Relatives by marriage have often been included with genetic relatives in the marriages prohibited by church and state. Sanctions against relatives by marriage might have some sociological significance but no biological significance, since the biological effect of consanguinity is to produce, in full view in the offspring, those recessive hidden traits coming from some of the ancestors common to the consanguineous parents. The previously hidden recessive traits may be beneficial or detrimental. They are likely to be detrimental more frequently than beneficial; this is because recessive mutations usually result in a less efficient enzyme rather than an improved system. Thus when recessive genes that have been carried along in the hidden heterozygous condition become homozygous, there is a small chance of improvement in the offspring but a large chance of a defect and disaster. Parents do not want to take such chances but often do not understand the genetic situation at all. This is especially true in cases of incest where the father-daughter or brother-sister union resulted from psychologically unhealthy family situations, oftentimes the female being mentally retarded and thus less able to prevent improper advances. A stepfather-stepdaugher relationship would be considered incestuous but is not consanguineous – as suggested to me by Dr. Carl J. Witkop, Jr.

What are the biological results of consanguineous unions? Adams and Neel [1967] found that 6 of 18 offspring of brother-sister or father-daugher unions had died or had major defects on follow-up six months after birth. No child had died or had a gross malformation in a control group. Seemanova [1971] studied a series of 161 children resulting from incestuous matings and compared them with 95 children from unions of the same mothers with unrelated partners. The mothers were mentally retarded in 14% of the cases. Recently, fewer incest children have been born as the mothers usually prefer to have a therapeutic abortion; marriages of fathers and daughters or brothers and sisters are prohibited. Some of the mental retardation in the children may be attributable to that of their parents, which probably played a role in the occurrence of the incest. In the Czechoslovakian sample there were 88 boys, five were stillborn or died of prematurity and eight more had died by the time of examination. Of the 72 girls, all were liveborn and nine had died by the time of examination. Thus there were 22 deaths among the 161 children resulting from incestuous unions. There was one child of indeterminate sex who died five days after birth, so there were 23 deaths, or 14.3% of the total. Of the 130 living incestuous children, after a few exclusions, there were 53 abnormals, or 41%. Thus there were 76 dead or abnormal children out of the 161, or 45.3%. The 95 half siblings without incest had only seven deaths and four abnormals, or a total of 11 deaths and abnormals or 11.6%. Finally, 40 of the 161 offspring of incestuous parents, or 25%, were severely mentally retarded while none of the 95 non-incestuous half siblings were severely retarded.

Incest is the closest union that can occur in man and obviously results in disaster for the offspring in from 33.3%–45.3% of the cases, using the data from the two studies just cited. The risk due to deleterious recessive genes is expected to be four times greater than for the offspring of first-cousin unions. Expectations from first-cousin unions would therefore be from 8.3%–11.3%, but this is too high because the incestuous parents are themselves defective more often than would be the case for first cousins.

If a person has a rare recessive gene in the heterozygous condition, the chance is 1 in 8 that the same gene will be present in any first cousin. For second cousins the chance of both carrying the gene is only 1 in 32, and for third cousins it is only 1 in 128. This last frequency of the gene would be no higher than that in the general population for most rare recessive genes. That is why there is little additional genetic risk for the offspring of distant relatives compared with nonrelatives.

The studies of first cousin unions show that the mortality rates and the frequency of abnormalities in the offspring are about twice those for the children of nonrelatives. In genetic counseling it is appropriate to give values of about 6% for mortality, and gross abnormalities for offspring of first cousins compared with those of nonrelatives of about 3%. The 6% is somewhat lower than the calculations given above of 8.3% for the reasons mentioned. The risks for children of uncle-niece, aunt-nephew, and other marriages more closely related than first cousins would be about twice the 6% risk for first cousins and thus fairly dangerous. Children from unions intermediate between second and first cousins would be at risks of about 4%–5%.

The genetic counselor is not likely to contact brothers and sisters contemplating parenthood but would certainly discourage them from such a dangerous undertaking. The situation is different for first cousins and more distantly related couples. In the latter situation the counselor can point out that society disapproves in general but that in specific cases it might be acceptable to take the increased risks and that the decision belongs to the couple.

Second cousins or more distant relatives can marry with the somewhat begrudging approval of society and their friends. It is true that the genetic risks are not significantly greater than for nonrelatives. However, it is true that a significant percentage of the children who are homozygous for any rare recessive trait have consanguineous parents. There is a biological relationship, known as Lenz's law, which states that the more rare a recessive gene in the population, the higher the rate of consanguineous marriages among the parents of the affected children. The law is valid because very rare recessive genes, which are concealed in the whole population, will seldom be found in both members of a couple. If, however, the parents are blood relatives, there will be a reasonable chance that both will carry at least one of the rare deleterious recessives possessed by one of their ancestors in common. Consequently, the more rare the recessive trait, the greater the likelihood that the homozygous child had consanguineous parents.

Fraser and Biddle [1976] presented some interesting results from genetic counseling cases seen in Montreal. They started with 58 probands with a disorder of genetic interest and whose parents were first cousins. A second sample of 27 probands with a genetic disorder and whose parents were second cousins was also obtained with appropriate control samples for both sets of probands. They then ascertained the number of affected siblings who had *different* genetic defects from those in the probands. They concluded that the risk of first-cousin parents having a child with a different recessively inherited disease from that of the proband child is low; that is, less than 1%. Infant deaths were about twice as high for the children of the two groups of relatives as for the controls (9.0% versus 3.5%). They concluded that increased infant mortality may result in part from environmental differences between consanguineous and non-consanguineous unions.

The physician can take on an aspect of the magician when a child is born with a defect due to the homozygosity of a rare recessive gene. He can point out to the parents that they frequently turn out to be distant relatives. Table 10-1 shows some rough data for children whose parents were found to be first cousins. The percentages would be higher if second cousins and all other consanguineous unions had been added.

The frequency of first cousin marriages in the United States is of the order of 0.5%, which is the base line for the data in Table 10-1. The inhabitants of small geographic isolates, such as island groups, are all likely to be related to each other in some way, even though the marriages of close relatives have been prevented by social pressures. This is the case for some small religious groups in the United States, such as the Amish. Interesting recessive traits have been found in such groups. These traits might not have appeared were it not for the obligate consanguinity that resulted from the religious and reproductive isolation from the surrounding communities of people of other faiths.

The physician may be asked about the laws concerning consanguineous marriages in the state where the prospective marriage would take place. An interesting treatment of all degrees of relationship is that of Farrow and Juberg [1969], with tables for each of the 50 states, the District of Columbia, and two territories. Fewer than one half of the states prohibit marriages with affinous persons, that is, in-laws. Such prohibitions are not due to genetic relationships, of course, but to legal reasons, such as inheritance of property and other social considerations. A little more than half of the states prohibit marriages between first cousins, with fewer and fewer states prohibiting marriages between the more distant relatives. Only one state, Oklahoma, prohibits the marriage of second cousins. All states permit more distant relatives to marry. Each state has its own laws about consanguineous and affinous marriages, which lead to an amazingly large number of different prohibitions for the whole country.

The birth of children with a homozygous defect brought about by consanguinity results in the elimination of two of these genes in each case, if the child fails to reproduce later on because of the disadvantage due to the genetic defect. This might be thought of as a beneficial way of improving the genotype of the

TABLE 10-1. The Extremely High Rates of First Cousin Unions Among the Parents of Children With Some of the Rare Recessive Diseases

Disease	Percent first cousin unions among the parents
True microcephaly	54
Wilson's disease	47
Ichthyosis congenita	40
Alcaptonuria	33
Xeroderma pigmentosa	26
Albinism	10
Phenylketonuria	10

species were it not for the heavy burden it puts upon the affected individual and his family. If people would refrain from unions with their blood relatives, the number of defective offspring each generation would be reduced by a significant percentage. This would permit a very slow increase in the frequency of recessive genes, as new mutations would continue to occur and elimination of them would be less frequent. However, one can hope that medical advances will continue so that the birth of defective children can be prevented and, failing that, treatments will become available that will "cure" the person of the defect. There are some philosophic individuals, apparently, who think that genetic disasters strengthen the character of the affected person and bring the family members closer together. Such advice, even if valid, always seems to be for the benefit of others, not for themselves. My impression is that the birth of an affected child leads to increased stress — and often to divorce — rather than to increase marital stability.

11
Blood Genetics

Death is the price of ignorance in blood group genetics! The technician can match blood types without genetics, but the results are precisely determined by the genes. More is known about the genetics of blood than about any other human tissue. This is partly because many chemical properties of blood display simple Mendelian patterns of inheritance. It is interesting that while predictions about the heredity of the blood groups are extremely reliable, it is also true that the *mature* red blood cells have no power of reproduction, as they have no nuclei present. The biological story is that the immature red blood cell did have genes present that controlled the production of the specific antigens that are found on the red blood cell as long as it exists. We have the unusual genetic advantage that probably there are not many chemical steps between the gene and the antigen in blood cells.

A-B-O and the Rhesus System

A person does not have antibodies in his blood that can agglutinate his own antigens, but everyone has all the rest of the antibodies of the A-B-O system. The AB person has both A and B antigens and, consequently, no corresponding antibodies. If he had such antibodies, he would be self-eliminating. As he has no such antibodies lurking about, he can receive red blood cells from any other type of A-B-O person and is therefore called a "universal recipient." The group O individual has all the antibodies and will agglutinate blood from all A-B-O types except his own. He is called a "universal donor" because his red blood cells have no ABO antigens and therefore cannot be agglutinated by ABO antibodies. The transfusion of O blood introduces into the host all the antibodies there are in the A-B-O series, which, one would think, should agglutinate some of the recipient's remaining precious red cells. Why don't they? The answer is that agglutination is a cooperative effect in a way. If the antibodies are too dilute, no agglutination occurs at all. Consequently, when the antibodies are slowly released into the patient's rapidly moving blood stream, the antibodies are instantaneously diluted below the minimum strength for agglutination.

The A-B-O system is not a rare trait. Everyone has at least one of the many alleles which compose the system. No normal person has more than two of these alleles as there are, of course, only two chromosomes in a pair. One of the major concepts of population genetics is that when there are two or more alleles of one gene locus all commonly present in a species, there must be an equilibrium among the alleles, which is maintained by natural selection. The major problem with this concept is that it is seldom possible to discover what factors operate as the forces of natural selection, or even that selection is occurring. The frequencies of the A-B-O alleles vary geographically, the frequency of people with blood group B being about 8% in the English and about 28% in Chinese people. Is it just chance that the B blood group is three times as frequent in the Chinese as in the English? There isn't any question but that random or genetic drift accounts for blood group frequencies found in many genetic isolates such as various Jewish groups. Changes in gene frequencies in relatively small isolates will be much more rapid than those which might be the result of natural selection. There is more information on the world distribution of all the human blood groups than any one physician would want to know in the 1,055-page volume by Mourant et al [1976].

It has been recognized for some time that there is negative selection against offspring who carry A-B-O or Rh antigens not present in their mothers. Presumably the mothers develop sufficient antibodies against the antigens in such offspring as to cause weakness or death of the offspring. The offspring receive the incompatible antigens from their fathers, of course. Probably the most extensive and reliable study of this topic is that of Cohen and Sayre [1968]. They found that there is a marked difference between A-B-O and Rh incompatibility in time of occurrence of fetal wastage in white mothers. A-B-O incompatibility is associated with a significant increase in early fetal deaths (15%–42% of them, depending upon the age of the mother) without any increase in late fetal loss. Rh incompatibility results in as much as a threefold increase in late fetal deaths without any effect on fetal loss at early gestational stages.

Levine [1958] was the first to point out that A-B-O incompatible matings sometimes provide protection from Rh erythroblastosis. Whatever the details of the interaction may be, Cohen and Sayre [1968] confirmed that double incompatibility (A-B-O and Rh) affords a lower risk of Rh selection in terms of late fetal loss than Rh incompatibility alone and, possibly, also a lower risk of A-B-O selection in terms of early fetal loss than A-B-O incompatibility alone. Further studies are needed for precise quantitation of these findings.

It is well known that 15% of United States white women are rhesus negative and 85% of those rh negative women who get pregnant would have an Rh positive mate. These unions have a substantial chance of producing a baby with erythroblastosis. Roughly speaking, about 1 in 200 of all white pregnancies resulted in an erythroblastotic baby at the time the first edition of this book was published

[1955]. At that time these babies usually died and were a very important source of wasted pregnancies. Fortunately, this is no longer true.

On October 1, 1973, an addition to the Maternal and Child Health portion of the New York State Sanitary Code went into effect. It reads in part,

"(a) It shall be the duty of the physician attending a pregnant woman to take or cause to be taken a sample of her blood at the time of first examination or as soon thereafter as practical. If the sample has not been taken and the pregnancy terminates as a result of an emergency, it shall be taken at the time of termination of the pregnancy. Such sample shall be submitted to an approved laboratory for the determination of blood group and Rh type.

(b) It shall further be the duty of the attending physician to evaluate every such patient for risk of sensitization to Rho (D) antigen and if the use of Rh immune globulin is indicated, and the patient consents, to cause an appropriate dosage thereof to be administered to her as soon as possible within 72 hours after delivery or other termination of pregnancy, whether induced or spontaneous."

The purpose of this legalism is to make sure that the 15% of rhesus negative mothers are detected before any harm can occur to the fetus as a result of the proliferation of antibodies by the mother and then to administer immune globulin to protect subsequent children from erythroblastosis.

In the early days before the globulin treatment was devised, pregnancies could be monitored for the presence of rapidly increasing antibodies in the pregnant mother and the obstetrician would remove the infant by Caesarean section about three weeks before term if that seemed warranted. There was a new danger introduced for the infant by prematurity, and fortunately such action is no longer necessary. It is an interesting observation that amniocentesis, which is now so important in prenatal diagnosis for many traits, was developed first for monitoring for erythroblastosis late in pregnancy. The deeper the yellow color of the amniotic fluid the greater the severity of the disease. The technological advances that have so greatly diminished the disaster of erythroblastosis fill one of the brightest pages of medical science. The world is indebted to many scientists for these advances, and first among them are the names of Landsteiner, Levine, and Wiener.

When the complicated genetics of the rhesus system was first developed, there was conflict over the terminology to be used. It was the blood of the Rhesus monkeys which was used to immunize rabbits first, while people were studied later. Wiener developed a system of nomenclature using the rhesus symbol Rh as the basis for the antigen with variations of the symbol Rh indicating other mutants of the basic gene, which were discovered throughout the world. The more logical CDE formulae produced by Sir Ronald Fisher have been combined with the Rh symbols in many laboratories to prevent confusion and errors which might result because of the two ways of naming the same antigens. It is not necessary to use space here to describe the two taxonomic rhesus systems, since all laboratories which do blood groupings are well aware of the situation and have appropriate charts which show the two systems in juxtaposition.

Cautions to the physician from the genetic counselor are in the area of awareness. A woman can have rhesus antibodies present before her first pregnancy and men can have rhesus antibodies present as a result of previous emergency transfusions. These are rare situations but they must be anticipated and proper action taken to prevent disasters. Always seek more information particularly in the extraordinarily complicated area of blood group genetics.

Abnormal Hemoglobins

Blood group genetics is only part of the subject of blood genetics. Blood group genetics concerns the antigens located on the surface of the red cells, while mutations affecting the structure of the most important part of the cell, the hemoglobin, work from within and early in the development of the red blood cell.

The most famous of the hemoglobin mutants is the one resulting in sickle cell anemia. The central atom of hemoglobin is the iron, which combines reversibly with oxygen, but the surrounding molecules of porphyrin and protein determine the precise conditions under which the iron takes up and gives off oxygen. The amino acid sequences in the α and β chains are controlled by genes in the nucleus of the developing red blood cell, and it is mutations in these genes that are transmitted from parent to child through the eggs and sperms. Most of the inherited abnormal hemoglobins are extremely rare, and only four of them, Hemoglobins C, D, E, and S are frequent enough to be of interest to the practicing physician. Of these, Hemoglobin S or sickle cell hemoglobin, was the first to be discovered in 1949 by Pauling. Persons who are homozygous for the S gene have their normal adult hemoglobin replaced by the abnormal type. This abnormal hemoglobin is precipitated in the form of angular crystals which distort the red cells into the sickle shape in which they are unduly susceptible to mechanical damage and destruction. Heterozygotes, with one gene for S and one for the normal Hemoglobin A, have both types of hemoglobin, through more A than S. Nearly all homozygous S children died of sickle cell anemia in the past, but in medically advanced communities many are surviving into adult life and are having children.

The heterozygotes for the sickle cell gene suffer no significant disabilities under normal conditions, though they do react unfavorably to anoxia because their cells sickle when there is low oxygen tension. The cells of the carrier will sickle on a sealed slide with cover slip, and thus screening could be done for the heterozygotes. Hemoglobin S is found mainly in blacks in tropical Africa or their descendants. Some cases are also found in Southwest Asia and India or wherever malignant tertian malaria has been endemic.

Sickle cell anemia is of interest to physicians who see black children because in this country about 1 in 400 of them can be expected to be homozygous for the S gene. It has been estimated that there are at least 50,000 persons in the United States who have the disease, and about two million others are heterozygous for the gene and have sickle cell trait. Thus about 1 in 400 blacks have the damaging disease, while about 1 in 10 blacks carry the gene for it but are not affected themselves. About 1 in 100 black couples would be expected to have the gene in both members of the couple and therefore be subject to the 25% Mendelian risk that each child would have the anemia and have a difficult life as long as it survived. The affected child becomes pale, tires easily, eats poorly, and may complain of pain in the arms, back, and abdomen. The first symptoms appear at 2–4 years of age, but occasionally earlier. In patients with sickle cell disease, the red cell survival is only 15–20 days compared with 120 days in normal persons. Despite this extreme hemolysis, the patient usually maintains a hemoglobin of 5.5–9.5 gm/100 ml by increasing red cell production five- to eightfold. An excellent description of the sickle cell disease crises was provided by Pearson and Diamond [1971].

Various publications are available, which cover the sickle cell story and are written for the general public. One of the most popular of them is the pamphlet by R.B. Scott, MD, and A.D. Kessler, MD, entitled *Sickle Cell Anemia and Your Child*, obtainable from the Department of Pediatrics, Howard University College of Medicine, Washington, D.C. 20001.

A disease often fatal in youngsters and affecting 50,000 persons is not a triviality in any sense. How did the gene for sickle cell disease achieve a frequency of about 5% in American blacks, which is very high for a lethal gene? The frequency of carriers of the gene in some parts of West Africa may be about 40%, a really astonishing situation. It was found that the reason for the phenomenon is that in Africa persons with the sickle cell gene in the heterozygous condition were protected to some extent from malaria. Thus the gene in the people it saved from death due to malaria was favored by natural selection. Some of the persons without a sickle cell gene would succumb to malaria and all of the sickle cell homozygous children would have died of the sickle cell disease in the olden times. So the carriers of the gene survived better than the two homozygous groups and increased the frequency of the gene generation after generation. The blacks in this country no longer need the protection of the sickle cell gene as malaria is not prevalent here. Consequently, the frequency of the sickle cell gene is decreasing gradually, as it is no longer advantageous in the heterozygous state and is very disadvantageous in the homozygous condition.

Sickle cell disease is an obvious candidate for screening programs, as the heterozygous person can be detected easily and inexpensively. There are sociological and psychological pitfalls to be avoided. Blacks are not pleased to find one more undesirable way in which they differ from whites by carrying the sickle cell gene. However, black leaders realize the damage done by the gene in the homozygous condition and have cooperated in the passing of state laws for the identification of carriers. Unfortunately, in some states these laws proposed that school children be screened. Screening black school children is a mistake in my opinion. Everyone who has reared children knows that they are often very illogical and even mistaken in their attitudes about health problems, and can visualize dangers where they don't exist. The 10% who would find themselves to be carriers might imagine all sorts of ills and difficulties related to being heterozygotes. It is hard to see how selective testing in integrated schools could avoid being psychologically traumatic for those tested. It would be foolish to test the white children. The only acceptable system, in my opinion, is the provision of ready access to testing facilities for blacks who are entering the reproductive age, perhaps a year or so before marriage. This doesn't seem to a highly efficient resolution of the problem, but most problems do not have completely satisfactory solutions.

However, great scientific advances at the most sophisticated levels have come to the rescue. If conception has already occurred and the parents are discovered to both be carriers of the sickle cell gene, a prenatal diagnosis can be carried out in one of two ways. Ultrasound and fetoscopy may be used to obtain a sample of blood from the fetus, but there is about a 5% chance that the fetus will be damaged or aborted. The second method uses the endonuclease, polyacrylamide gel, autoradiography method to study the hemoglobin chains and does not use the more dangerous fetoscopy. This latter method seems to hold the greater promise for detecting the homozygous sickle cell or thalassemia fetus and permits appropriate utilization of the therapeutic abortion, if requested. This whole area is clearly that of the specialist, and couples with both members carrying the trait should be referred to the nearest genetics center where the proper specialist can be contacted. See Panny et al [1979].

Abnormal Hemoglobins C, D, and E are less important than Hemoglobin S for various reasons that can be omitted here. The thalassemias are also related to malaria resistance in that the original selection for the heterozygotes was in malarial areas. In this country most often the affected people are of Mediterranean origin. The counseling and screening problems are similar to those for sickle cell anemia.

The readiness with which changes in hemoglobin structure are observed and the high likelihood that changes in structure will have clinical consequences have made mutations at the hemoglobin loci among the best known of all human gene loci. A large number of hemoglobin variants are known, and the chemical analysis of the amino acid substitutions and deletions can be carried out with great precision and relative ease. Most of the mutations will never be seen by any one person and need not be listed here.

Genetic counseling for carriers of the sickle cell or thalassemia genes as they are discovered in screening programs as a routine part of health care brings up ethical questions as to how to approach the person with the trait. Numerous questions as to privacy and sensitivity arise. Rowley et al [1979] have provided a sophisticated report of their experiences with β-thalassemia in an unsuspecting population in Rochester, New York. Anyone concerned with carriers of hemoglobin anomalies should read this report of what transpired at the Rochester HMO.

There are hereditary non-hemolytic anemias, but these are of low incidence in the population and must be omitted here.

Lymphocytes — Leukocytes

The lymphocytes are specific types of white cells, which carry antigens that are also present in other tissues of the body. If a person is to donate an organ for transplantation, his or her lymphocytes will be crossmatched with the serum of the recipient in an effort to detect antibodies in the recipient that are incompatible with donor antigens. Such preformed antibodies could be responsible for rejection of the donor organ. The specific antigens possessed by the donor are determined by using the lymphocytes as the source of the antigens. These antigens are histocompatibility antigens and are called the HLA system. The HL means 'human leukocyte' and the A means that it was the first gene locus designated. The terminology has changed over the years with recent substantial recodification provided in Table 1 of the monograph on the subject edited by Bodmer [1978]. The components of the HLA system reside at four gene loci in the order DBCA on chromosome 6 with many alleles at each of the four gene loci. The four loci are tightly linked. All the antigens are inherited as simple co-dominant characters, that is, each allele can be detected if present in a person as can A and B of the more familiar red cell groups.

The HLA system is now the most complex genetic system known in man. Its great importance has been understood only during the last 10 years. It should be the most useful system for paternity determinations as soon as laboratories are available throughout the country, which can administer the tests. Its utility for organ transplants stimulated much of the genetic work in the area, while the greatest value of the system may result from its association with diseases such as psoriasis, hay fever, Addison's disease, celiac disease, juvenile type diabetes (see Chapter 16), ankylosing spondylitis, and others. A vast amount of work remains to be done in confirming, extending, and exploring these associations between the array of HLA antigens and the appropriate diseases. These associations of the HLA with specific diseases could be of use in genetic counseling where they are very strong, but would not be helpful where they are weak. Svejgaard et al [1975] showed that in nine studies totaling 445 patients with ankylosing spondylitis, there were 88% with HLA-B27 and only 8.6% of the controls had HLA-B27. Thus relatives with HLA-B27 present would have a better than 80% risk of developing the disease, under appropriate conditions.

Pregnancy is nature's transplantation experiment, and it is not known why the HLA "incompatible" fetus is tolerated without signs of rejection for nine months. Apparently the HLA antigens are present on fetal cells very early. The complexity of the system would indicate that practically every fetus would have some HLA gene arrangement which would be incompatible with the mother, so some compensatory arrangement must be present in order to have healthy babies. Whatever the reason for this favorable outcome, few genetic counseling cases will result directly from HLA incompatibilities. The system will be of great use in clinical medicine, paternity determinations, and twin diagnoses, all areas in which the geneticist will be deeply involved. All HLA phenotypes are individually rare, and it is difficult to find two unrelated individuals who have identical HLA phenotypes; thus the positive determination of paternity and the diagnosis of twins should be entirely feasible. Svejgaard et al [1975] state that no other genetic system in man shows a similarly high degree of polymorphism and that there is a theoretical probability of finding about 20 million different phenotypes in the population — obviously, a most exciting area for human geneticist! Those interested in this area should study also the papers by Bach and van Rood [1976].

Other Systems

There are numerous other systems of considerable significance, but many of them are individually rare and thus cannot be taken up with the thoroughness they deserve.

Hemophilia is one of the best known of the rare blood disorders. It is the "royal" disease, which was carried by Queen Victoria and transmitted to various royal houses via her daughters. The consequences were serious, particularly in Russia and Spain. There are several other recessive "factor" deficiencies that result in faulty blood coagulation.

Antibodies are globulins, chemically, and most of them move in an electric field with the gamma globulin fraction. The area of immunoglobin differences is too complicated to consider here and does not give rise to genetic counseling problems as such. There is variation of the gamma globulin molecules, which results in an elaborate system of alleles at what is called the Gm locus. These are of great interest to the geneticist because they contribute to the possible different phenotypes, which allow paternity determinations and the diagnosis of zygosity in twins.

The haptoglobins were the earliest known and most extensively studied of the serum protein polymorphisms. These proteins are alpha-2-globulins and have the property of binding hemoglobin. At least two gene loci are involved with this system, with various alleles at each locus. The system is of interest to geneticists as a model for reconstructing details of gene evolution, and it probably has important associations with various diseases still remaining to be found. However, the system is not of interest for genetic counseling purposes yet. An elegant review of the subject is that of Sutton [1970].

A final topic in the blood genetics are of interest for genetic counseling is a fairly frequent disease, pernicious anemia. There was a great deal of interest in the disease at the time of the discovery of the vitamin B12 therapy. Since that there has been less interest, presumably because the therapy is quite successful. The genetics of the trait probably belongs to the multigenic category, as it is about twice as common in females as in males. About 10% of the first degree relatives develop the condition. In addition another 10% will develop a hypochromic anemia. Thus there is a moderate risk, but only a moderate burden, as a result of medical progress. Stevenson and Davison [1970] present an excellent brief discussion of the hereditary disorders of the blood.

This chapter does not exhaust the wealth of different blood systems known. However, it considers those likely to present themselves for genetic counseling. In many of the polymorphic blood systems there is no "normal" phenotype. Blood group B is presumably as normal as group A. It is the transmission and recombination of the different antigens in various family relationships that result in medical problems and in requests for genetic counseling, hopefully before the medical problem has developed.

12
Disputed Paternity

Illegitimate births are one of the major problems of American sociology. Illegitimacy results from irresponsible behavior in most cases and strains the tolerance of the rest of the population, which has to help shoulder the resulting burden. Children born into one-parent families have a much greater chance of delinquency than those with both parents present. It is fortunate that the prejudice against the illegitimate child now is mild compared with the old puritanical spite and vengefulness. These are not value judgments, but self-evident truths accepted by every sociologist.

Berkov and Sklar [1975] reported that in 1973, in New York State, about 9% of births to white mothers were illegitimate and about 45% of births to nonwhite mothers were illegitimate. Sing et al [1971] reported that from 1.5—3.5% of Caucasian children in the United States born in wedlock were estimated, on the basis of genetic markers, to show discrepancies between legal and biological parentage.

Paternity suits are not filed by poor blacks as a rule. The *publicized* paternity suits are against movie stars such as Charles Chaplin and Marlon Brando. The latter was acquitted in court, though both were eliminated as the biological parent as a result of negative blood tests. At the time of Chaplin's trial, evidence from blood tests was not admitted in California courts; it is now admissible. Large amounts of money are involved in the paternity cases of celebrities, whereas other types of cases, such as possible swapped babies in hospitals, have a greater emotional involvement. In the latter cases maternity may be questioned. The number of cases in which genetic parenthood needs to be determined increases all the time, and the courts are gradually becoming more receptive to scientific evidence for such determinations. Courts will now admit blood group evidence if it excludes paternity, but no case is known to me where a man was adjudicated to *be* the father of a child as a result of scientific tests.

In Germany and Denmark the public welfare system is less permissive about child support and attempts to determine paternity in a much higher proportion of cases than is customary in the United States. Blood group evidence in those countries is acceptable in determining that a person is, indeed, the father of a given child.

Chakraborty et al [1974] listed 57 immunological and biochemical genetic systems for which testing could easily become routine in many United States laboratories. They included the probabilities for black, white, and Japanese of excluding a wrongfully accused male if the latter is randomly selected from the general population. There is always the danger of mistakes in classifying blood groups and other biochemical systems, and this should always be foremost in one's mind before making any paternity decision. Exclusions should be based on more than one blood group or biochemical system. However, the accuracy of these scientific tests is at least equal to other lines of evidence for judicial consideration.

The authors cited above show that the probability of excluding a falsely accused white male from paternity with at least two exclusions, is 96% and for a black male is 93%. Thus a man who has been falsely accused of paternity has an excellent chance of being absolved of responsibility for the conception of the child. If all possible fathers of the child are tested, it should be possible to exclude all but one of them. This would be practically certain if the HLA system is included.

Polesky and Krause [1976] showed that 93.3% of falsely accused white men could be excluded if only seven blood group systems were used, that is, ABO, Rh, MNSs, Kell, Duffy, Kidd, and HLA. They point out that the new HEW regulations state that, "The IV-D agency shall identify laboratories within the State which perform legally and medically acceptable tests, including blood tests, which tend to identify the father or exclude the alleged father from paternity. A list of such laboratories shall be available to appropriate courts and law enforcement officials, and to the public upon request."

The next question is whether the person who cannot be excluded is really the father of the child. Can paternity be attributed to any male who is not excluded on the basis of laboratory tests? The answer is yes, if the child possessed an extremely rare inherited biochemical variant (say one with a frequency of 1/50,000 persons) and this variant was present in the putative father but not the mother (see Chakraborty et al [1974]). Can the same degree of certainty grow out of a collection of polymorphic genetic systems such as are available in various laboratories? The above authors think that such a case can be made. It depends upon the types of testing that a particular laboratory is equipped to make. Bias [1975] pointed out that fewer than 57 loci are usually tested for. She also notes that when any reasonable battery of tests is accepted by those concerned, the case is most likely to be settled out of court.

How small does the probability of *not* being the real father have to become before we can decide that the person *is* the biological father? Langaney and Pison [1975] claim that there is no practical way in which this discrimination can be made. They insist that multiple exclusion with safe genetic markers is the only way at present that paternity diagnosis can be done with accuracy. Their view is the conservative, and probably correct, answer for 18 or less polymorphic systems. However, a practical battery of tests probably will be devised in the future, which can be passed only by the biological father — or his identical twin! In the meantime, the model builder should prepare simulated tests that would demonstrate what would be needed for paternity diagnosis as a feasible technique. This test would include the extremely important HLA system as mentioned in Chapter 11 of this book.

There is a pressing need for much more research on paternity diagnosis, as I am willing to predict that there will be a change in welfare policies in the United States in the near future. We probably will insist that efforts be made to force working males to support their illegitimate children instead of leaving it for the taxpayers to do, as is so often the case at present. A good place to start one's literature search before embarking on any research in the field is the comprehensive bibliography compiled by Koch [1976] and the book by Sussman [1976].

Finally, the knowledge and techniques are already available so that wrongly accused males can be excluded by means of the usual serological and biochemical systems, including the HLA group in almost 100% of the cases. In addition, paternity can be positively established in many cases by the presence of the same rare blood group haplotype in the biological father and the child. The major deficiencies at present are the financial mechanism to pay for the extensive tests and the skilled personnel properly established in convenient laboratory settings to do the work.

After all this consideration of the use of blood groups as the primary way of determining the paternity of a child, it may turn out that some other method will become the primary tool, and the elaborate blood group determinations will become only additional or secondary supporting material. I refer specifically to the banding techniques which permit the identification of marker chromosomes. It is true that chromosome 1 of both man and gibbon have had about the same banding and morphology for some 50 million years, as described in Chapter 2; it is also true that each person shows remarkable polymorphisms in the chromosome banding picture. Hecht [1971] suggested that, except for identical twins, each person's karyotype might prove to differ from everyone else's, perhaps being the cytological equivalent of the fingerprint. The study of Müller et al [1975] showed that in a newborn population no two individuals with identical Q- or C-band patterns were found. Friedrich et al [1977] used the banding technique to show that a pair of male twins were dizygotic rather than monozygotic. This was of interest because only one of the twins had Down syndrome. See also Verma et al [1977] and Van Dyke et al [1977].

The chromosome polymorphisms should work equally well in resolving paternity disputes as in determining twin zygosity. Any strikingly different-looking chromosome present in both baby and one of the putative fathers would be a positive determination, while additional chromosome correlations of this sort would clinch the case. Inclusive paternity determinations are now both feasible and appropriate.

13
Cystic Fibrosis — A Challenge

The reader may wonder why a whole chapter should be devoted to only one disease. There are several answers to the question. Cystic fibrosis (CF) is the most frequent serious disease with a simple Mendelian recessive basis among white children in the United States; thus, it is of great practical importance. We still know very little about the molecular biology of the disorder. It is still impossible to screen for carriers of the recessive gene with sufficient accuracy. We do not know whether the gene (or genes) for cystic fibrosis became so frequent because of some type of advantage possessed by the carrier of the gene compared with the homozygous normal person or as the result of genetic drift. Is genetic heterogeneity an important factor in the situation? All these unknowns, and others, provide us with a sharp challenge to do more significant research to resolve some of the questions which are so obvious and which have defied resolution for a relatively long time.

The first statistical analysis showing that cystic fibrosis was a Mendelian recessive trait was done by Andersen and Hodges [1946]. In the paper by Goodman and Reed [1952] the frequency of the disorder was estimated to be from 0.7 to 1.0 per thousand births. This may have been too high, though the figure given by Bowman and Mangos [1976] of 1 in 1,600 or 0.4 per thousand is not greatly different, considering the difficulties of establishing this particular diagnosis at birth. The center of origin of the high frequency of the gene must have been somewhere in northwestern Europe. Of people descended from ancestors born in that area, about 1 in 20 will be a carrier of the recessive gene, assuming a frequency of affected births of 1 in 1,600. There are many millions of carriers of a cystic fibrosis gene in the United States, and there are probably one hundred thousand or more couples with both members heterozygous but not aware of it. Many of these couples will escape having an affected child.

I have no data as to the number of children born each year with cystic fibrosis, or with the potentiality of developing it. However, my estimate is that there are some 2,500 births each year in the United States in which the disease will be diagnosed at some time (1 in 1,600 frequency, 4,000,000 births annually). This is a disease of great significance, particularly when it is remembered that the cost of treating a single patient may be a financial disaster for the family. Clearly, this disease is one where screening of conjugal couples to determine whether both are carriers would be of great help.

The most practical "screening" test is naturally that of finding the affected homozygous patient as soon after birth as possible. The wide spectrum of symptoms which vary from patient to patient probably means that the basic defect is a number of metabolic steps prior to the observed pathology. Consequently, it will be more difficult to find a screening test useful for all patients. The tests should be capable of detecting in the newborn one or more aspects of chronic pulmonary disease, pancreatic exocrine deficiency, and abnormally high sodium and chloride concentrations in sweat. The disease escapes detection in some patients until young adulthood, and late diagnosis seems to become more, rather than less, common.

The sweat electrolyte abnormality is present from birth and persists throughout life. The sweat test, when properly performed, is accepted as the simplest and most reliable laboratory procedure for confirming the diagnosis of CF in children. The method of choice seems to be that of Gibson-Cooke which is described in Appendix B of the NAS/NRC Committee Report [1976] on evaluation of testing for cystic fibrosis. The sweat test is not as definitive in persons over 18 as it is in children. Enzyme assay will be especially useful for adults as these patients may be tall, well nourished, and relatively healthy looking. Not only tryptic activity but activity of chymotrypsin and carboxypeptidase are absent or greatly reduced in CF.

The most practical test for widespread screening of the *newborn* is the meconium test. All patients with meconium ileus and about 80% of all infants born with cystic fibrosis have abnormally high concentrations of albumin in their meconiums; see Stephan et al [1975]. According to Warren J. Warwick, MD, in a personal communication, with newborn infants having the genetics for cystic fibrosis, one would detect 80% of them but would miss the 20% who would be false negatives. No doubt many more than 20% are missed and die without diagnosis under present procedures. Consequently, newborn testing would be very useful for these cases. However, there would be about 1,200 false positives found for every 80 bona fide cases. These 1,200 babies would all have to be given the sweat test at about 4–6 weeks of age in order to eliminate the approximately 1,180 false positives and to find the remaining 20 real cases. In the meantime the 1,180 pairs of parents have been subjected to needless worry which is ethically undesirable.

Except for the above complication there is no ethical objection to screening for the homozygous CF patient because treatment may be helpful when a patient is detected, that is, diagnosed. Early diagnosis is not only humane but a basic concept of preventive medicine. It is a matter of medical and financial priorities as to whether widespread screening for CF is desirable. The priority for widespread screening seems to be low, but this may be due more to human lethargy than to scientific logic. One major barrier to widespread screening is that the sweat test is not successful with newborn babies. For infants it is still expensive and not foolproof. Much research remains to be done before there can be a really satisfactory test for CF.

There is still no test that will detect cystic fibrosis in utero. However, screening of the children from proved carrier couples is certainly indicated. These heterozygous parents will know that they have a 25% chance of having an affected child and will want to have a determination of the genotype of the child in order that they can plan for future years.

Final diagnosis for the homozygous child is still expensive and complicated. It should be done in specialized centers in order to increase accuracy. There are now 121 CF centers distributed throughout the United States (see the NAS/NCR report, 1976), so it should be possible for the general physician to obtain satisfactory testing for his patients. If a simplified test becomes available, the physician's office will become the test location with increasing frequency.

If testing for the homozygous CF child is still cumbersome, then the detection of the carrier person is downright discouraging. A practical test for the carrier is needed badly. The sweat test would seem to be the most promising way of detecting heterozygous persons, but it suffers from poor discriminating power; values for the three genotypes overlap with each other. Another test which has caused considerable excitement results from the observations that homozygotes and heterozygotes have a factor in their sera that inhibits the motion of the cilia in oyster gills and rabbit trachea. But it is difficult to control laboratory variation in the ciliary activity, and the evaluation of results is too subjective to be widely applicable. There must be a better way found to assay for the CF factor. While ciliary bioassays are still unfit for clinical separation of the three genotypes (homozygotes, heterozygotes, and normals), they remain the only monitoring device to follow purification procedures for isolation of the cystic fibrosis mucoinhibitor, as Bowman et al [1977] point out.

The positive aspect of screening is to separate those people who are free from any genes for CF from the carriers and the patients. It should be remembered that in any normal control group we would expect as many as 5% of the persons to be heterozygotes. As one would expect, testing for normality in regard to CF is a discouraging proposition. However, from a practical point of view it could be of great value because persons with low sodium and chloride levels would seldom be carriers of the CF gene and thus would not be expected to have affected children. Thus, many relatives of a CF patient could be relieved of apprehension concerning their own reproduction. Those with average and high levels would still have to bear the anxieties present before the test, but they also might feel somewhat relieved in that they would now have a friendly physician interested in their problem, who would provide genetic counseling for them.

There is no question but that people with similar problems profit from joining together. Parents of CF patients through their local and national organizations not only have been mutually helpful but have provided millions of dollars for research on this most challenging disease. The chronicity and often fatal outcome of CF are sources of severe psychological stress for patients and their families. The study by Boyle et al [1976] showed that even in patients with good daily physical performance the emotional adjustment was often impaired. Intelligence is unaffected in CF patients, but in the psychological area there cannot help but be difficulties because of the expectation of premature death and the sterility problem, especially for males. Naturally, the physician is going to be one of the most important factors in the emotional adjustments of patients and parents. Any physician who has a CF patient or involved family should study the Boyle et al paper carefully, as the subject is much too extensive to be developed here.

However, the viewpoint of the cystic fibrosis patient should not be too pessimistic. The study by Shwachman et al [1977] of 70 patients who were over 25 years of age showed that 26 of the males and 15 of the females were college graduates; there was one MD and one Associate Professor of Psychology already. None of the males had children, but 18 of the women, married for a total of 70 years, had 7 children.

One of the most thrilling sagas of triumph over a handicap just unfolded at the 1980 Winter Olympics. The flag bearer for the United States athletes was Scott Hamilton, a young man only five feet, three inches tall and weighing 110 pounds. His growth was arrested between the ages of 5 and 9 years because of a bout with Shwachman's disease, a disease similar to CF. He was selected as flag bearer because of his outgoing personality and for having qualified for the International Olympics by winning the bronze medal in the U.S. men's figure skating contest. His dazzling performance in the long Olympic figure skating program gave him a fifth place in this most prestigious of athletic happenings. He met the challenge of a most debilitating disease, and his triumph should be an encouragement to all CF patients.

The challenge of cystic fibrosis extends even to its genetic basis. In the paper by Lowe, May, and Reed [1949], my commitment to a Mendelian recessive basis for the disease was unequivocal. Since then it has been obvious that genetic theory has become more complicated and we realize that genetic heterogeneity for any trait is practically a requirement as a result of the mutability of the billions of genes present in the population of the world. Presumably, new mutations to genes for cystic fibrosis occur all the time and they may be slightly different in their effects. Those new mutations which occur at the same gene locus will presumably interact with each other, when in the same child, to produce a CF patient. If two mutations were at different gene loci, then the child might not be affected, or less severely affected, than usual. One of the handles we have on this type of problem is a determination of the frequency of CF in the first cousins of a patient. Each uncle or aunt of the patient has a 50% chance of being a heterozygote for the CF gene, and the spouse of the uncle or aunt about a 1 in 20 chance of carrying the same gene or a similar mutation. The couple have a 1 in 4 chance of producing an affected child, if both are carriers. The product of these probabilities is 1 chance in 160 that any first cousin will be affected. Actually, the chance is less than this because the trait is not diagnosed early in life in all cases. This important piece of research was carried out by Danks et al [1965]. Their results were in excellent agreement with the simple concept that only one gene locus was of importance. There are statistical biases in all such studies, and Danks points out one of his; in those families with a child still attending the hospital, the segregation ratio was well above 1 in 4, but it was very low in those families which proved to be most difficult to locate. In all probability there will be an excess of families with several affected children in most samples because of the greater likelihood that a family will come to the attention of specialists when more than one child is affected. Oddly enough, researchers seldom have mentioned the probable presence of this bias in their samples, which rather uniformly include a few too many affected siblings. If one is willing to accept the likelihood of such probable statistical biases, there is little evidence from family studies that mutations for cystic fibrosis occur at more than one gene locus in any significant frequency. However, Schaap et al [1976] have been able to provide a two-gene-locus model in which the double heterozygote (AaBb) is the affected person, while all the other relevant genotypes result in phenotypically normal individuals.

Danes et al [1977] contributed to the genetic challenge of cystic fibrosis by studying a family in which the one affected member was assumed to be heterozygous for different alleles or for different genes at two independent loci. Their concept is that the atypical (mild) clinical features of this adult CF patient were due to two different CF genes combining to produce a genetic compound expressing a mild form of CF. This is an entirely different model from that provided above by Shaap et al.

It is hard to evaluate the importance of these suggestions that two independent gene loci are required to produce the CF patient. The papers of Wright and Morton [1968] and Conneally et al [1973] provided large amounts of data showing excellent agreement with the segregation ratio expected for a single gene locus. Their segregation data strongly suggest complete penetrance of the homozygote in the families they located.

The greatest challenge to the geneticist is that of explaining the high frequency of CF patients, which may be as high as 1 in 1,600 births, to couples of northwestern European descent when the disease has been practically lethal in the past. It is difficult to explain this high frequency as a direct relationship to the mutation rate known in man. Genetic drift is difficult to accept because of the high frequency of the "lethal" gene in millions of Caucasians. It would require that the carriers of this gene all be descendants of a few persons who were the original mutants, or even one original mutant. This can happen in small populations over relatively few generations. However, we are concerned with millions of Europeans and their descendants over hundreds of generations. Another alternative is that the mutant gene in the heterozygous condition gave a selective advantage to the carrier persons living in conditions peculiar to Caucasians for many generations in the past. Under other conditions the frequency of the gene would be correlated with its mutation rate. Wright and Morton [1968] suggested that the typical Caucasian allele may arise so rarely by mutation as not to have been available for positive selection in non-Caucasian populations. This suggestion implies that the typical Caucasian allele must have mutated at about the time of the differentiation of the Caucasian race from the other races. Burdick [1977] has pointed out that the reproductive advantage of the cystic fibrosis heterozygote need be only 2.63% to maintain the present gene frequency.

This means that a gene, such as that for cystic fibrosis, which has a high frequency in a large population and is lethal in the homozygous condition must be maintained by either a high mutation rate or confer some selective advantage upon the heterozygotes. In this case it seems more reasonable to assume that there has been an advantage to the heterozygote at some time in the past which might be still effective.

Genetic counseling for cystic fibrosis can be carried out as for any simple recessive trait. There are additional complications which the counselor will have to deal with. The parents of the CF baby will keep pressing for more opinions as to how long their child can be expected to live. They certainly will be distraught as to how the hospital expenses can be met. The interest in this disease has led to widespread reports in the media, and the CF parents' clubs have garnered whatever information is available. Consequently, the parents may ask for information about the sterility that affects about 95% of CF males who survive to become adults. Edwards [1973] has pointed out that counselors and parents' groups can increase anxiety for the couple and even for the general population if the available information is handled in an incompetent fashion.

There are still physicians who are not aware that there is a genetic basis for cystic fibrosis in spite of the widespread publicity about the disease. Furthermore, much of the genetic counseling which is done is carried out in an inappropriate clinical environment by clinicians with insufficient training in human genetics. Under such circumstances it is quite possible that the parents of the CF baby will reject the information about genetics which is presented to them on a very casual basis. While they should not be "bugged" by emphasis on their situation as obligate heterozygotes, they should be made clearly aware of their genotypes and the possible consequences thereof. Naturally, their instructions should not be punitive but supportive in the reality of their serious situation. Much greater efforts must be devoted to the education of physicians as to how to do genetic counseling for cystic fibrosis, as this challenging disease is one of the most important genetic pediatric problems in white children in the United States.

14
Twins

The arrival of twins is usually a matter of mild consternation, mixed with a tingle of anticipation as to their similarities or differences in development. Sometimes they are looked upon with pleasure as an omen of fertility and plenty. In Australia the aborigines used to liquidate one of the twins; twins meant double trouble to them. Their view was that a woman could hardly expect to carry more than one child in her arms, because she had to carry her nomadic husband's spears and other belongings as well.

In the United States, twin pairs appear in roughly 1 in 90 births, triplets in about 1 out of every 90 births squared, quadruplets in 1 out of every 90 births to the third power, and quintuplets in 1 out of every 90 births to the fourth power. Sextuplets are rare indeed and all six seldom survive much past birth. Thus, at least one in every 50 persons is, therefore, a member of a twin pair or other type of multiple birth.

Aside from their interest as curiosities, twins are most valuable scientific material. As everyone knows, there are two kinds of twins, identical and fraternal. The identical twins start off as one egg fertilized by one sperm; the early embryo then separates into two identical parts. As the identical twins are merely one individual walking around in two bodies, they must be genetically identical, gene for gene. Furthermore, they are forced into very similar environments, so that it is perhaps odd that they show any differences at all. Fraternal twins start as two separate eggs fertilized by separate sperm. They are no more alike, genetically, than ordinary brothers and sisters. They do share an environment that is more similar than that of ordinary brothers and sisters. There is no evidence that this somewhat greater similarity in environment makes a very profound difference in the expression of their measurable traits.

If all human traits behaved in the clear-cut mendelian fashion that albinism and Huntington's disease do, twin studies would not be necessary as an aid in unraveling the complications that the environment often superimposes upon a mendelian pattern of heredity. There seems to be a rule that the most common defects have the largest environmental component, which makes it difficult to tell whether the hereditary basis, for the abnormality is a simple dominant or recessive. The environment acts to suppress the expression of the abnormal gene in some cases but fails to do so in others. It can be appreciated that this unpredictable behavior of environmental factors would upset the classic orderly mendelian ratios, especially if they are the complicated ones that result when more than one gene pair is involved.

Pairs of twins are useful in detecting the relative effects of heredity and environment upon the expression of a disease or trait. If a trait is highly hereditary, both members of a pair of identical twins will be expected to show the trait. If one identical twin shows the trait and the other member of the pair does not, the disagreement must be due to environmental differences between the two twins, because the genes of one are exactly the same as the genes of the other. Fraternal twins are no more alike genetically than ordinary siblings, so if they differ in that one twin has the abnormality and the other does not have it, one would not know whether it was the environment, the heredity, or both that differed for the two members of the pair. Actually, they differ in both their heredity and their environment. There are, of course, exceptions to these generalizations — but the general rules still hold.

We know that the genetics of the standard blood groups behaves according to the mendelian rules and that the environment has no significant effect on the kind of blood groups a person has. Consequently, with identical twins, both members of the pair will be expected to have exactly the same blood groups. In fact, it is from the blood groups found in a pair of twins that we decide whether they are identical or not. Many other characteristics are included in determining the zygosity of a pair of twins. The use of a battery of tests allows one to calculate the probability that a pair of twins are identical according to the method of S. M. Smith and Penrose [1955]. The necessary tests are expensive and for some kinds of work are unnecessary, because a simple visual classification of the twins may suffice. Photographs are not a good method for zygosity determination because they leave out so much of the personality of the person. Finally, if it is important to know whether twins are identical for such purposes as organ transplants, the Smith and Penrose type of zygosity determination should be carried out.

Interpretation of concordance rates of less than unity for identical twins has always perplexed human geneticists. One would not expect perfect concordance if the trait is variable in its expression or if it is often sporadic in its appearance. It is possible to calculate a valid heritability from twin data for any trait with an appreciable genetic basis. This was demonstrated by C. Smith [1974]. He showed that proband concordance rates in twins provide estimates of correlation among twins in liability to a trait. The correlations in turn can be interpreted genetically by the expression $2(r_{mz} - r_{dz})$, which estimates the coefficient of genetic determination for the trait. The physician need not employ this formula but can find the estimated heritability much easier by using the graph provided by Smith

[1970, p 89]. His graph will be referred to frequently in subsequent sections of this book.

What about the genetics of twinning itself? Two quite different phenomena are involved, identical twins being the products of a single fertilized egg, while the fraternal twins result from two different fertilized eggs. The hereditary mechanism for the monozygotic (MZ) twins should therefore be different from that for the dizygotic (DZ) twins. Monozygotic twinning might be considered a congenital malformation. Bulmer [1970] has restated the old speculation that a common mechanism may be responsible for both MZ twinning and some congenital malformations. He showed that monozygotic twinning is remarkably constant over a wide range of conditions, and there does not seem to be any specific genotype responsible for it. It is perhaps a response to a rather specific environmental condition at the time of fertilization or soon thereafter. This may be a delay in development of the egg but the reasons for that are unknown.

Dizygotic twinning is a form of superovulation. Gemzell and Roos [1966] treated about one hundred women who had long-lasting amenorrhea with injections of gonadotropin. Forty-three pregnancies resulted, of which 20 were of single children, 14 twins, 2 triplets, 3 quadruplets, 1 quintuplet, 2 sextuplets, and 1 septuplet, although many of the higher multiple births aborted. The trade-off here is between the risk of continuing infertility or superovulation. The results were obviously a response to the hormone injections. Hormone levels of the mother are presumably under genetic control to some degree. Thus the heredity of the mother may be of significance in the production of dizygotic twins. It should be remembered that about half of her heredity for twin production came from her father. In view of the long term pressure of natural selection we may assume that the genetic basis for gonadotropin levels is of the multigenic type.

A neat little discovery by Jovanovic et al [1977] showed that it is possible to detect twin pregnancies with certainty by 40 days after the last menstrual period of the mother. This is a simple and practical finding which helps provide for optimal care in twin pregnancy. Women who are carrying twins will have human chorionic gonadotropin levels at least twice as high as those of women with singletons. There seems to be no overlapping of the values between the twin and singleton pregnancies, especially as late as 70 days after the last menstrual period. It is unusual to have any test which discriminates so clearly between two groups of individuals. The prognosis for twin pregnancy can be considerably improved by this early test, which also is the test for pregnancy itself, and allows the adoption of prophylactic measures at an early date to ensure better health for the twins.

Among white twins roughly one third are monozygotic and the other two thirds are dizygotic. Among blacks the proportions are about 29% monozygotic and 71% dizygotic. Myrianthopoulos [1975] showed that fetal and neonatal deaths amounted to 17.3% and were significantly higher among like-sexed than unlike-sexed pairs. The difference was contributed almost entirely by deaths in male pairs. In the majority of cases both members of a pair died. The most common cause of neonatal twin deaths was respiratory distress syndrome. In his study of the large material in the Collaborative Perinatal Project it was found that 24.1% of the monozygotic twin children had a malformation, while only 14.8% of the dizygotic twin children had a malformation. The latter figure was not statistically different from the 15.6% of malformations in the singletons. All these results seem to be shockingly high. The reason for this is that about half of the malformations are classified as minor. The significant excess of malformations was found only among the identical twins for both the major and minor malformations. The above data are of high quality and can be used with confidence.

In the Myrianthopoulos study the most frequent major malformations are those of the alimentary and musculoskeletal systems followed by those of the cardiovascular and central nervous system. Specifically, the highly significant excesses of major malformations in twins were for encephalocele, tracheoesophageal fistula, malrotation of the alimentary tract, and inguinal hernia.

There has always been a popular belief that twins are not quite as bright as singletons and there are, unfortunately, some data which confirm the notion. The difference is partly due to the prenatal environment, for twins brought up as singletons, still perform at the intelligence level of twins and not of singletons. It may also be due partly to the higher incidence of congenital malformations in twins, especially those of the central nervous system. Fortunately, the performance of twins, relative to that of singletons, tends to improve, at least from four to seven years. This suggests that prematurity is also a contributing factor and that its detrimental effects may be reversible. The slightly poorer intellectual performance of twins compared with singletons should be considered to be a difference but not a detriment in the total picture, as there are other differences which give substantial advantages to twins, such as their companionship with each other.

An interesting demonstration of the really touching relationships that develop among twins is that of Taylor [1970] who found 50 marriages of twin pairs with twin pairs. Only one of these resulted in divorce. All but one of the 49 stable marriages found the twin foursome living close by; three quarters of them shared the same house even after there were children. The children usually considered both pairs of twins to be their parents even though they had no trouble distinguishing between them.

A seldom explored but important aspect of twin studies has been introduced by the genius of Scarr-Salapatek (unpublished). She pointed out that in 18—40% of twin pairs, one or both members were mistaken about their zygosity. Parents hold more similar expectations for their identical than their fraternal twins. The evidence of greater environmental similarity for MZ than DZ twins is overwhelming.

Presumably the identical twins contribute to the similarity of their environments to a large extent by their choices which result from their identical genotypes. What effects result from the greater similarities in the environments of the MZ twins compared with the DZ twins, and how significant are they? The pairs of twins who are mistaken about their zygosity should be helpful in such investigations. It should be borne in mind that some pairs of fraternal twins are more alike genetically than others because of the random nature of genetic segregation.

If genetic similarity is the major determinant of behavioral likeness, then DZ twins who believe themselves to be MZs may be no more alike than other DZs: and MZs who mistake themselves for DZs may be no more different than other MZs. If, on the other hand, beliefs about zygosity influence the extent to which co-twins are behaviorally similar, then DZ twins who think they are MZs should be as similar as true MZs. Likewise, MZs who believe they are DZs may be as different as true DZs. Actual zygosity was determined by concordance or discordance at 12 to 23 loci by the Minneapolis War Memorial Blood Bank.

Scarr-Salapatek and her associates found that on intellectual measures, co-twins resembled each other according to their true, rather than perceived, zygosity. Blood group similarities at 12 loci showed that those DZ twins who believed they were MZs were actually more similar genetically than the other DZ pairs. She found that perceived zygosity is not an important bias in studies of genetic variance in intellectual skills. For personality variables, perceived zygosity may have some effect on fraternal pairs who believe themselves to be identical.

The comfortable assumption of approximately equal environmental variance for MZ and DZ twins was shown to be tenable. The fact that MZ twins generally experience more similar environments seems to result from their genetic identities. Thus, twin studies as a whole provide insight into the relative importance of genetic and environmental factors in the variation of many of the traits which are of importance in the practice of genetic counseling.

Twin studies of personality traits, such as the pioneering study of Gottesman [1963], were based on samples too small to give reproducible results, though they pointed to future expectations for this exciting area of research. The recent book by Loehlin and Nichols [1976] on heredity, environment, and personality of 850 sets of twins showed that, "identical twins correlate about 0.20 higher than fraternal twins and it doesn't much matter what you measure..' Some traits give higher correlations than others but the difference between the two kinds of twins always seems to be about the same regardless of the nature of the trait. It has been known for many years that random factors are important components in the variance of embryonic development of laboratory animals. The above authors had the same difficulty in trying to pin down specific environmental factors related to personality traits. They concluded that the environment works in "remarkably mysterious ways," at least as we conceive of its effects at present.

It is certainly true at present that our psychological and personality tests are not constructed in the way we would like to answer present day questions. However, twin studies are likely to be the best technique for utilizing whatever tests there are in attempting to find out more about the nature of these mysterious environmental factors.

15
Multifactorial Inheritance

The next chapter (16) is composed of a number of sections, each dealing with one of the frequent traits which the physician is likely to encounter in his practice. They have many characteristics in common and all are subject to environmental factors which, in toto, may have a greater effect on the expression of the trait than the necessary genotype. That is, there is a necessary genotype required for any expression of the trait, but the genotype is not sufficient to ensure the production of the trait; environmental factors of various kinds contribute to the phenotype which we observe. They are quasi-continuous traits.

Geneticists classify differences between individuals with respect to a given trait as *discontinuous* variations (in which the trait is clearly distinguished from the normal state and usually fits a Mendelian ratio) or *continuous* variations (in which there is no sharp division between normal and abnormal and each individual can be placed on a scale depending on many genetic and environmental factors, each with a small effect). Traits in a third category, *quasi-continuous* variations, seem to be discontinuous but behave genetically as if they result from a continuous distribution of an underlying variable, separated into the normal and abnormal classes by a threshold. The rest of this chapter will be an effort to demonstrate the way in which these quasi-continuous, or threshold, traits behave. It should be useful to review a little of the literature on this subject because the quasi-continuous traits such as congenital hip disease, pyloric stenosis, diabetes, and so on are the largest group for which genetic counseling is requested, and they are the most difficult group to understand from a genetic viewpoint.

A multifactorial model of the etiology of a trait where the genetic and environmental factors are so numerous, but individually so weak, results in a normal curve of liability for expression of the trait. The aim of understanding the genetic basis of diseases of the multifactorial type is partly that of assessing the pliability of the disease in response to environmental variation, which could be imposed either for prevention or for therapy.

The general concept that applies to most of the common malformations was demonstrated by Carter [1965]. He pointed out, for instance, that for a malformation with an incidence of 1 in 1,000, a monozygotic twin concordance of only 40% is compatible with a heritability of 80%, but cautioned that heritabilities calculated from twin data may not be very reliable. He also pointed out that for a multigenic trait with a population incidence of about one per thousand (and a heritability approaching 100%), we would expect a frequency of affected persons of about 3.5% in the first degree relatives, 0.7% in the second degree relatives, and 0.3% in the third degree relatives. There is a sharp drop in incidence between the first and second degree relatives of about five times (3.5%–0.7%), while the drop in incidence from second to third degree relatives is much less. If we were dealing with a single dominant gene with reduced penetrance, we would expect a drop of one half between each degree of relationship, that is 3.5%, 1.7%, and 0.8% instead of the 3.5%, 0.7%, and 0.3% shown above.

It might seem that it would be easy to distinguish between multigenic inheritance and Mendelian inheritance of a single dominant gene with lack of penetrance. However, the comparisons in the case of a specific trait never come out neatly enough to be convincing, and Smith [1971] has shown by elaborate computer methods why a distinction between these two types of inheritance for a specific trait will usually be impossible. This leaves us with the necessity of using empiric risk data for genetic counseling without knowing what the genetic mechanism responsible for the trait may be in such cases.

However, it is possible to estimate the heritability for the trait so that we have some idea as to the pliability of the disease, that is, the degree to which environmental manipulation might be possible for prevention or cure of the disease.

Perhaps the most practical method for obtaining a heritability estimate is that developed by Falconer [1965]. His method is primarily a device for converting the information contained in the incidences of the trait in the relatives into an estimate of the correlations between them. The correlation between the relatives for a trait is proportional to the heritability of the trait. Said in another way, the heritability expresses the extent to which the phenotypes exhibited by the parents are transmitted to their offspring.

As mentioned before, we assume a gradation of factors causing the trait. All persons above a certain point on the scale, the threshold, will show the trait, and those below the threshold will not. The term liability includes not only the individual's genetic susceptibilities, but also the whole combination of circumstances that make the person more or less likely to develop the disease.

For the quantitative development of the theory for quasi-continuous traits it is necessary to define the variation of liability as being in the form of a normal curve. This gives a unit for the expression of the degree of liability, the unit being the standard deviation. The threshold, the point on the scale of liability beyond which all individuals are affected with the trait, provides a fixed point by which to compare different populations or groups of relatives with different incidences of the trait.

Falconer [1965] demonstrated that one could estimate a heritability for a trait in a specific group of relatives with no further data than the incidence of the trait in the general population (the frequency of the probands) and the incidence in a specified group of relatives of interest. It is a simple matter to read off an estimate of the heritability of the trait from the now famous graph available in his 1965 paper. Furthermore, he illustrated the method with four practical examples which are reproduced below:

	Heritability
Renal stone disease	46 ± 9%
Congenital pyloric stenosis	79 ± 5%
Clubfoot	70 ± 8%
Peptic ulcer	37 ± 6%

Any method that is so easy to apply could be easy to misinterpret and misuse, because such an easy procedure invites one to forget the possible sources of error which are inherent in most any technique used for an estimation of values. For instance, the liability may not be a continuous distribution. That is, there might be a fairly important dominant gene involved in, say, pyloric stenosis, along with numerous additive genes. A much greater sin has been that of comparing the heritabilities for different races of people. A particular heritability value is a property only of the population it was derived for and cannot be extrapolated to some other unrelated population. A much more deceptive problem with heritability values is that they include the within-family environmental variance as well as the genetic variance. It is this within-family environmental variance which we are most anxious to separate out, but so far it has not been possible to do that.

The impossibility of partitioning out the within-family environmental influences means that heritability values would tend to be too high rather than too low. This will be particularly true when the regression is on the second member of a pair of identical twins because of the attempts of parents to provide the same type of environment for the two members of the identical twin pair. Smith [1970] has revised Falconer's method to remove the two types of bias resulting from the skewedness of the distribution, first of the affected probands, and second of the affected relatives of the probands. Unfortunately, it is still impossible to remove the within-family environmental component included in the heritability estimate. A revised graph is provided by Smith to replace that of Falconer [1965], and an additional new graph allows one to obtain the heritability values when twin data are available.

Smith [1970] demonstrated that the heritability can be high with a low concordance between identical twins and, in fact, concordance between identical twins will be high only if the heritability is *very* high (ordinarily when the trait depends upon a single gene locus with complete penetrance). Exceptions would be high concordance for both identical and fraternal twins in their susceptibility to infectious diseases such as measles, where most everybody is susceptible to the virus.

The next principle which will be considered in this introduction to multigenic "curable" traits is the interesting deviation of some traits from the 1:1 sex ratio. Sometimes there is an excess of affected males for a trait like pyloric stenosis with a sex ratio of 5:1 or an excess of females with a sex ratio of about 1:5 for congenital hip disease. The deviation for most traits is not as extreme as the two examples given above, but is characteristic for the trait and is repeatable with each new survey. This has nothing to do with sex linkage in the geneticist's sense!

The liability for the trait, where there is a deviation from the 1:1 sex ratio, still has a bell-shaped normal distribution with the affected individuals to the right of a threshold. The threshold for the females is farther to the right than the threshold for the males when the trait has a higher prevalence in males. Therefore, affected females must be relatively farther out towards the tail of the distribution than affected males and would be genetically more predisposed to the trait. Consequently, a higher proportion of affected individuals is to be expected among the near relatives of affected females than among those of affected males. In principle, therefore, when malformations occur more often in one sex than in the other, the proportion of near relatives affected will be greater when the proband is of the less-often-affected sex. For a recent discussion, see Reich et al [1975].

An elegant example of the above principle is that provided by Carter and Evans [1969] for pyloric stenosis shown below:

Proband	Sons	Daughters	Brothers	Sisters
Male	19/347	8/337	21/546	15/565
640	5.5 %	2.4 %	3.8 %	2.7 %
Female	20/106	7/100	25/273	10/263
352	18.9 %	7.0 %	9.2 %	3.8 %

We see that the proportion of affected relatives is about three times greater for female probands than for male index cases. This sharp difference is readily explicable on the multigenic hypothesis but not on any other genetic hypothesis.

Still another principle of multigenic inheritance is that the incidence of affected offspring should increase after the proband has had one affected child. Using the Carter and Evans data again for pyloric stenosis we do find the expected increase in the frequencies of affected children: 3 out of 17 (17.6%) where the father was the proband and 5 out of 13 (38.5%) where the mother was the index case. These numbers are small, but they are very convincing evidence for a multigenic basis for the trait because no increase would be expected on any single gene hypothesis. That is, fathers would have been expected to produce only about 3.9% affected offspring and mothers 13.2% affected offspring subsequent to the first affected child on a single gene basis.

As a final aid to our consideration of multigenic inheritance, there is a paper by Bonaiti-Pellié and Smith [1974] which provides three tables of risk figures for use in genetic counseling. The three tables are for cleft lip ± cleft palate, pyloric stenosis and anencephaly-spina bifida. There are 27 rows of different family combinations for each of four possible parental types. The risk figures given are *not* observed empiric risk figures, but are those calculated on the hypothesis of multigenic inheritance based upon the incidences and correlations of liability appropriate for each of the three traits.

The three tables are helpful, as they cover the range of risks which one might expect to find for any multifactorial trait. The risks are uniformly low for the majority of probable counseling situations usually encountered.

Let me complete this introduction to the traits of most importance to the physician, because they are those he is likely to see, with a simple table which is part of Table 14–1 in the excellent text by Nora and Fraser [1974], which they adapted from Table 1 of Smith [1971]. My table, 15–1, gives the empiric risk figures for any trait which has a frequency of 1 case per 1,000 persons for heritability estimates of 100%, 80%, and 50%.

People who come for genetic counseling do not concern themselves with small differences between risk figures. They want to know whether their risks are high or low, so the counselor's problem is largely that of interpreting to them how low is low and how high is high – not an easy task! Nonetheless, the ability to demonstrate the relevance and meaning of a specific risk is one of the most important qualifications of a genetic counselor.

The genetic counselor must provide an understanding of the meaning of the risks of a repetition for a trait, and after that he should provide information, if needed, as to the size of the burden to the affected person and the relatives, as has been explained previously.

TABLE 15–1. Recurrence Risks (%) for Multifactorial Diseases According to Heritability and Number of Affected First-Degree Relatives for a Trait With a Frequency of About 1 in 1,000 Persons

	Affected Parents								
	0			1			2		
Heritability	Affected sibs			Affected sibs			Affected sibs		
(%)	0	1	2	0	1	2	0	1	2
100	0.1	4	11	5	16	26	62	63	64
80	0.1	3	10	4	14	23	60	61	62
50	0.1	1	3	1	3	7	7	11	15

16
Some Common Multifactorial Diseases

The previous chapter was written as an introduction to the multifactorial diseases which will be considered in the present chapter. These diseases or disorders are "common" in the sense that, taken together, they form the largest group the physician will encounter in which the treatment involves action beyond the administration of medications or other relatively simple ministrations. For all these diseases there will be questions about heredity and genetic counseling with which the physician should cope in a scientific fashion, either by knowing or obtaining the counseling data himself or by referring the patient to a counseling center.

Genetics has a poor image in the popular view and this is partly due to the erroneous notion that if a disease is genetic or hereditary nothing can be done to relieve the patient of the symptoms. There are very few diseases for which modern medicine can do nothing at all. To be sure, there are some inborn errors of metabolism such as Tay-Sach's disease where the patient still dies, but each of such lethal disorders has a very low frequency and they are seldom encountered by the family physician. Instead, he will be confronted with allergies, cancers, circulatory system problems, facial clefts, and the other multifactorial traits, some of which will be considered in this and other chapters. Most of these traits are treatable at least to a large degree. The result may not be perfect or complete but generally it is satisfactory to the patient. Consequently, genetic disorders, and all disorders are genetic to some degree, should be viewed as an intriguing challenge rather than rejected as a hopeless object for despair. To be sure, prevention is much better than any palliative but if prevention has failed, we are indeed fortunate to have the remedies which are being provided by medical research.

It is not the purpose of this book to describe medical treatments, but to provide the data for genetic counseling for both prevention and for the situations where prevention was impossible, as it still is impossible, for the bulk of the counseling cases where the couple had no way of anticipating that their child would be born with an anomaly. There are at least palliatives for all the traits described in this chapter and, with the addition of numerous, somewhat similar traits in subsequent chapters, it makes a long list of multifactorial traits. All of these traits are "constitutional" in the sense that numerous gene pairs are involved with each anomaly and presumably the gene pairs represent "normal" variants of the genotype in that some of the relevant gene partners may be present in substantial frequency throughout the whole population. We still know very little about such polymorphic gene loci. There is a great need for more research on the basic genetics of traits like pyloric stenosis for our theoretical and practical understanding of multifactorial traits as health problems. This research is of particular urgency in order that prenatal detection can be developed, which then provides avenues of prevention.

The reader is free to reject the idea that all or any of the traits included in this chapter are multigenic because it is impossible to *prove* that any of them are multigenic in the strictest sense. We cannot provide the mapping data to show where each of the many genes for any of these traits is located on its chromosome. However, every trait is genetic to some extent, though it may be a quite indirect relationship. Consequently, it should be possible to determine which genetic mechanism is of primary importance for each trait. Those placed in the multifactorial and multigenic categories arrive there more by a process of elimination than as a result of scientific proof. It is the choice dictated by ignorance but nonetheless the product of our best judgment, which stems from what we do know about each trait.

Even though we must wait for successful research and development, the sections of the present chapter should be useful for physicians because these are the traits that will be sprung upon them, usually without warning, and as they all have a genetic basis, confusing as it may be, genetic counseling will be required. The threats the patients will feel will usually be much greater than the statistical risks warrant. Consequently, the physician can achieve considerable satisfaction from reducing the anxieties of these very unhappy "patients," who are most often the normal parents of an abnormal child.

ALLERGIES

It is of alphabetical significançe only that the first topic to be considered in this extensive chapter with many subsections is that of the allergies. The problem with the genetics of the allergies is that everyone (almost) seems to have an allergic reaction against something. However, the main difficulty is not only the high frequency of allergies, but also the confusion as to their relationships with each other. It is a taxonomic or diagnostic problem that makes a genetic study very difficult and inconclusive.

The age of onset difficulty contributes further to the complexity of the picture. As a personal example, I was in my early sixties when I had my first contact dermatitis reaction to any substance, in this case a new variety of shaving lotion which was proven to be the offending agent by laboratory tests in the University dermatology department. It is said that I had severe eczema as a baby, but I do not know whether these two long-separated allergic reactions were related or not. My wife and children all have different allergies, my daughter having had severe asthma at one time.

It is conceivable that almost everyone has an allergic reaction to something at some time during his or her life. If this be the case, the genetic background for the host of different allergies must be heterogeneous indeed. According to McKee [1936], only 34.3% of a large and relatively unbiased population sample could be considered non-allergic. But this estimate is probably much too high.

The state of our knowledge about the genetics of allergies is still in such chaos that there isn't agreement yet as to the general magnitude of concordance for allergies in twins. Older studies showed a concordance of 88% for identical twins where both twins had an allergic reaction, though not necessarily to the same substance. The fraternal twins showed a concordance of 64%, which presumably indicates that both the heredity for allergies and the environmental factors are practically universal.

More recent studies such as that of Lubs [1972] present much lower concordance figures for twins. She considered previous studies to be biased upwards, but her material (the Swedish twin study of 7,000 pairs of adult twins) may not have been too comprehensive, as the data were very retrospective and obtained by mailed questionnaires. Nonetheless, except for contact dermatitis, the differences in concordance between the identical and fraternal twins were all statistically highly significant. For instance; for asthma, the identical twins were concordant in 19.0% of the pairs while both fraternal twins were asthmatic in only 4.8% of the pairs. For the more ubiquitous hay fever, 21.4% of the identical twin pairs were concordant, while 13.6% of the fraternal twin pairs had hay fever in both members.

The comparatively low values obtained by Lubs [1972] for the twin material also extend to her results for the first degree relatives of the twin proband. If the proband had asthma alone, the risk for the first degree relative to have asthma was 5.8%, while if the proband had asthma and either eczema or hay fever, the risk for asthma was 12.6% in the first degree relatives. Thus the heavier loading for allergies in the proband correlated with a higher frequency of asthma in the relatives.

The figures of Van Arsdel and Motulsky [1959] show that of the 16.7% of college students with asthma and/or hay fever, some 383 parents out of 1,942 (19.7%) had asthma and/or hay fever. This latter figure is higher than the 5.6% found in their control group. They also found that the presence of both asthma and hay fever in the same individual was associated with an especially high frequency of allergies in the family background. This would be evidence for a multigenic type of heredity.

Smith [1974] found a strong correlation between social class and skin test reactions to a variety of allergens. This is interesting because she found that the top social class had the greatest percentage of reactors to skin tests. For instance,

22.7% of the high-medium social class persons reacted to the house dust tests, while only 6.3% of the low-low social class persons reacted to it. She speculated that, "The clearly lower incidence of sensitization in the low-low income group suggests that social factors offer protection (perhaps somewhat as environmental and immunologic factors combine to reduce the incidence of paralytic polio-myelitis in poorer populations)." In this situation the bottom social class comes out on top of the good health picture.

One of the most important questions is whether a person inherits a general allergic predisposition or a predisposition for a specific allergic disease. Asthma and hay fever are associated more often than would be expected by chance according to their respective frequencies in the general population. Fortunately, no person is so unfortunate as to exhibit all the known allergies. Presumably there is some genetic specificity in the expression of the allergies, because agents such as strawberries, house dust, animal detritus, and other possible irritants must be encountered many times in every person's life. My guess as to the answer to the above question is that there are both general and specific genetic factors con-tributing to the predisposition to allergic reactions. This is a common sense deduc-tion and a rather discouraging one as it implies that it will be a long time before any of the numerous hypothetical gene loci will be identified or mapped on their respective chromosomes. In other words, the multigenic explanation for the total allergy picture seems to be the most likely answer. However, individual gene loci may be involved with specific allergies as well as a general diathesis for allergic reactions.

There has been considerable research on the immune reactions, indicating that such single loci are involved with specific antigens. For instance, the work of Levine et al [1972] showed that clinical ragweed pollenosis and IgE antibody production specific for antigen E (the major purified protein antigen from rag-weed pollen extract) correlated closely with the HLA haplotypes in successive generations of seven families selected for affected individuals. This implies dominant inheritance with low penetrance, which, however, cannot be distin-guished from multigenic inheritance.

In a recent study, Gerrard et al [1978] accepted the concept that high levels of serum IgE are frequently associated with allergies. Their path analysis gave a genetic heritability of 0.425 for serum IgE levels in a sample of white families. Their segregation analysis indicated, in addition to significant polygenic herit-ability, a major regulatory locus RE, with homozygotes re/re maintaining per-sistently high levels of IgE.

Regardless of any genetic explanation of the predisposition to allergic reactions, we can provide empiric risk figures for counseling purposes. All the figures are conservative, as none have been corrected for age of onset and especially do not include allergies other than those mentioned by the various authors.

The following data (Table 16-1) come from the paper by Van Arder and Motulsky [1959] for asthma and hay fever.

Somewhat lower empiric risk figures were provided for first degree relatives by Lubs [1972].

In summary, the situation is still fluid. Genetic counseling is rather academic for a trait such as an allergy because of the high frequency of the trait in the population and because an allergy is often almost a relative blessing compared with what other genetic defects one might have to suffer from. However, we do get rather frequent requests for genetic counseling for allergies and present the risks in the above table which can be used for first degree relatives.

Table 16-1. Risk Figures for Each Offspring of Parents of the Following Mating Types

Mating type of parents	Percentage frequency of each type	Percentage risk for each offspring
Positive x positive	0.8	58.1
Positive x negative	14.5	38.4
Negative x negative (neither parent with asthma or hay fever)	84.0	12.5

CANCERS

Cancers are second only to heart and circulatory system disorders as causes of death in the United States. We get few counseling requests at the Dight Institute regarding either cancers or circulatory disorders unless they affect young people. Everyone accepts the inevitability of death; consequently, disorders such as the major types of cancers which take their toll among adults do not usually initiate genetic counseling requests. However, some cancers do affect babies and young persons, and we do get some counseling requests for traits such as retinoblastoma and leukemia.

Due to the tremendous heterogeneity of tumors and cancers we can expect a great diversity of causes involved. Some may be associated with chromosome aneuploidy, others with single mutant genes; others may respond to viruses. But probably most can be best explained by a genetic-environmental interaction; that is, multigenic heredity which is expected to respond significantly to environmental factors and changes in them, which, of course, would include viruses and various chemical and physical agents. For a general review, see Anderson [1975].

In the Down syndrome, the risk of myelogenous leukemia is increased 30 times (to 1/95) over the population risk of childhood leukemia of 1/2,880 as stated by Nora and Fraser [1974]. There is no tidy explanation yet as to what produces this well established relationship between trisomy 21 (Down) and leukemia. Childhood leukemia not only appears in familial aggregations, but it also appears in "clusters" in specific geographic areas, and one "cluster" of adult cases of leukemia has been found. Obviously, searches were made to try to identify the environmental factors in the particular geographic location which triggered the development of the leukemia cases. Unfortunately, the search attempts have not been successful. This statement has to be qualified in one sense because it is well known that heavy doses of irradiation, such as the fall out from the atom bomb, resulted in many cases of leukemia a few years after the irradiation. For a comprehensive review of genetics in human cancer see Mulvihill et al [1977]. For the role of the chromosomes see the splendid book by Sandberg [1979].

Studies such as that of Levan and Mitelman [1975] indicate that the chromosome aberrations in cancer cells are not random. The nonrandom clustering of specific chromosomes as the probable basis for neoplasms has been revealed by the chromosome banding techniques developed in the last few years. Much research remains to be done in this area. At present the data suggest that the genetics of the host cell is a necessary factor in the development of essentially all forms of neoplasia. We still have little idea as to how the viruses, chemicals, physical trauma, or other agents trigger the chromosomal and gene reactions which result in the cellular disorganization called cancer.

The fifth edition of McKusick's [1978] catalog of traits lists over 50 tumors or cancers which have a single gene locus basis. These include among the dominant traits, neurofibromatosis, polyposis, retinoblastoma, and tuberous sclerosis. Among the recessive traits are Bloom's syndrome and xeroderma pigmentosa. The Wiskott-Aldrich syndrome is one of the well known X-linked traits which can be included here. All of these are rare traits and should cause no great difficulty in genetic counseling because of their simple genetics and reasonably good penetrance.

The genetic counseling difficulties come with the common types of cancers. The major common types of cancers seem to be genetically distinct from each other and there are probably genetically distinct subtypes within the major groups. Some of the subtypes might depend upon single gene loci for their expression, such as Gardner's polyposis, but usually there is little evidence of single loci when one makes allowance for coincidence in the "loaded" pedigrees. We will look at the counseling risks for a few of the major types of cancers at this point.

About 20% of deaths are from some form of cancer. Thus many families will be expected in which numerous members will have cancers on any theory of causation, just as a result of statistical coincidence. Of course there was some definite causation in each individual case, but it didn't have to be the same cause for all the affected persons. Our counseling cases are often members of these highly visible families with more than their share of affected relatives. It should be pointed out to them that the risk of anyone selected at random dying of cancer is about one fifth and that, furthermore, many cancer cases die from some completely independent cause not related to their cancer. One can see already that their risk would seldom be much higher than the 20% for the general population. In fact, for a specific type of common cancer, the risk to a relative should be substantially less than 20%.

Cancers of Specific Sites

1. **Breast.** Twin data are surprisingly scarce for the various cancers. In the Hauge et al [1968] compilation of 4,368 same-sexed pairs of the Danish Twin Register, there were only 4 pairs out of 23 identical twins pairs which were concordant (17.4%). There were 6 concordant pairs of fraternal twins in 47 pairs (12.6%). All the other types of cancers listed gave lower concordance percentages or none at all. Obviously, the heritabilities would be low with such low (or zero) agreements between identical twins; the concordance for fraternal twins were as high as those for the identical twins for some of the types of cancer.

In view of the above, there is little evidence for strong heredity of the common cancers. The correlation with aneuploidy in the cancerous cells is strong evidence of a genetic basis for the cancers, but not evidence for the transmission of genes for cancers from one generation to the next, as is, on the other hand, the case for Mendelian dominant traits like neurofibromatosis and retinoblastoma. However, there is good evidence for weak heredity of a predisposition to breast cancer.

Anderson [1972] showed that in patients with premenopausal diagnoses, the frequency of breast cancer in their relatives was 6.7%, which was three times higher than the 2.3% in similar-aged control relatives. No significant increase in frequency was observed for relatives of patients with postmenopausal disease compared with controls. He also showed that when the disease in the patient was bilateral, the frequency in the relatives increased to 13%, and if both premenopausal and bilateral, the frequency was 17%. The frequency in relatives of patients with unilateral disease, whether pre- or postmenopausal, was 3.4% – only slightly higher than control values. These data are in good accord with expectations for multigenic traits with "high" frequency in a population. They do not suggest dominant inheritance, which some authors have confused with the coincidental aggregation of affected relatives.

Anderson et al [1958] showed that the likelihood of a female first degree relative of a proband developing a breast cancer by age 85 is about 10%. In practice, the chances are lower than this because few counseling clients live to be 85 years old.

2. Colon or rectum. The most interesting cancers (genetically) in this group are the polyposis types such as Gardner's syndrome and the Peutz-Jegher syndrome. All the polyposis types of cancers seem to be inherited as Mendelian dominant traits with fairly high penetrance. The geneticist would be correct in predicting that these are relatively rare in the population; they are found to be present in about 1 in 10,000–20,000 persons.

Cancers of the large intestine, without polyposis, are frequent and are an important cause of death. The twin data from Hauge et al [1968] show very low concordance, 7% for identicals and 4% for fraternal pairs. The numbers are small but nonetheless indicate that heredity has a weak role in intestinal cancer, as is the case for all the frequent cancers. This stands in contrast to the striking Mendelian dominant inheritance found for the rare syndrome types.

Carcinomas of the colon or rectum are frequent in whites, intermediate in orientals, and relatively rare in blacks. There is no established difference in the sex ratio of affected persons with cancers of the colon.

There is a familial predispostion to intestinal cancers, which seems to be stronger than one might expect from the low concordance of the identical twins. The data agree with those for other common multigenic traits when the frequencies of affected first degree relatives are considered. There have been three competent studies carried out at about the same time, ie, those of Macklin [1960], Moertel et al [1958], and Woolf [1958], on first degree relatives with appropriate controls. These studies showed that some 4%–10% of the first degree relatives had cancers of the large intestine, about three times the rate found in the controls. For the most severely affected probands the frequency among first degree relatives with colonic cancers could be as high as 24%.

3. Stomach. One could expect an excess of stomach cancers among the relatives of probands with intestinal cancers and vice versa. This is not the case, the etiology of each location in the digestive tract being independent of the others. Once again we find a very low concordance for identical twins of about 7% (see Hauge [1968]), while the fraternal twins had a concordance of only 2%, which is no higher than the expectation for the general population. The incidence of stomach cancer varies considerably between countries, being relatively low in whites in the United States compared with orientals. It is commoner in north Wales than in southeast England. There is some strange association between blood group A persons and stomach cancer. Group A individuals are about 20% more liable than people of other blood groups to have stomach cancer. This is not a genetic linkage, but an interesting physiological association which still needs experimental explanation.

There have been several family studies of stomach cancer that provided the now familiar results that about 4%–10% of the first degree relatives also had stomach cancer, which is about three times as high as the frequency in controls and the general population. With respect to the frequency of cancer as a whole, most of the studies did not reveal any significant difference between the first degree relatives of stomach cancer patients and the controls.

Woolf [1961] found that although blood relatives of probands had twice as much gastric cancer as their controls, spouses of the probands did not differ from their controls. This suggests that important environmental factors for stomach cancer are not operating in the environment shared by the spouses, so they must be experienced only by the relative or else genetic factors account for the difference in frequency of stomach cancer between the relatives and their spouses.

4. Lung. At the beginning of the century lung cancer was a relatively rare type of cancer. Now it is the greatest single cause of cancer death in the United States. The rise in the frequency of lung cancer is correlated with the rise in cigarette smoking during the last few generations.

Tokuhata [1964] demonstrated a familial factor in lung cancer which is rather small but about the same in men and women. The effect of cigarette smoking was greater than that of the familial factor in smokers. Furthermore, there was a synergistic interaction between the familial and smoking factors. Those who had both characteristics were subjected to a remarkably increased "booster-effect" in the pathogenesis of the disease. Those who possess both characteristics are exposed to a fourteenfold greater risk than those without either characteristic, that is, those without a family history and who have not smoked cigarettes. Even with cigarette smoking, the risk of lung cancer in the first degree relatives, after an age correction had been added, would probably not be greater than 5%. This is my speculation, as Tokuhata did not attempt an age correction.

The environmental agent of greatest importance for lung cancer is clearly cigarette smoking. Any success in reducing cigarette smoking would decrease the frequency of lung cancer more effectively than any eugenic measures which could be attempted.

5. Skin. Basal cell carcinomas are the most frequent type of skin cancer followed by squamous cell carcinomas and finally by malignant melanomas and other rare types. Prolonged exposure to sunlight, ionizing radiations, and chemicals seem to be the most important environmental agents.

There are the usual Mendelian dominant types of syndromes such as the nevoid basal cell carcinoma syndrome (see Howell and Anderson [1972]). This not only behaves as a clear-cut dominant trait, but it also has 95% penetrance. However, most cases of skin cancer have few or no affected relatives, and their relatives will have a low risk of skin cancer. The relatives have a much higher risk of dying of some other entirely independent type of cancer. We should remind ourselves that 20% of the population dies from some type of cancer and an additional large percentage survives the cancer to die from some other cause.

There are no data known to me for the bulk of the skin cancer cases which presumably are of the multigenic type. However, we can expect the risk to first degree relatives of the proband to be low, certainly not more than 5%, while if there is an affected first degree relative in addition to the proband, the risk could be doubled to 10%. The risk of 10% is not high when one thinks of the nearly 50% risk for the dominant syndrome types of cancers. Clearly, the correct diagnosis as to the type of cancer in the proband is of the greatest importance as the counseling risks are so different for the multigenic predispostion types compared with the dominant syndrome types.

6. Other cancers. It would not be useful to continue our listing of types of cancers and their genetic predispositions because the situation does not change significantly for any of them. There are the identifiable rare syndromes with Mendelian dominant heredity with fair to excellent penetrance, but the majority of cases behave like traits which are frequent in the population and depend upon the multigenic "constitution" or predisposition of the person in their reactions to specific carcinogens. As just mentioned, the risk of a repetition in the first degree relatives for the same type of cancer as exhibited by the proband is small, but the risk for some other genetically independent and unrelated type of cancer is about 20%.

In summary, the genetic counseling center will see the occasional case of retinoblastoma, neurofibromatosis, and some of the other Mendelian dominant, recessive, or X-linked syndromes with neoplasms. The center will have few requests about breast cancer, stomach cancer, or other common types of cancers. The family physician will probably have little likelihood of receiving requests for counseling for the syndromes with clear-cut Mendelian genetics because they are so rare, but probably he will be asked about the heredity of the common cancers. Fortunately, the risk to a first degree relative of a common cancer proband is around 5% and around 10% if there has been a second relative with the same type of cancer. The risk of having any kind of cancer for anyone is better than 20%, as cancers are the second most common cause of death in the general population of the United States. While genetic counseling may be useful for an understanding of the cancer picture as a whole, and for the Mendelian syndromes with high risk, it loses some of its utility for the individual common cancers for which the risk of repetition is low. Of course, the knowledge that the risk of a repetition of the same type of cancer is small is good news and will be helpful to the counselee.

We have the peculiar situation that a counselee has a greater chance of dying from a different kind of cancer than the specific type which had appeared in his family and which he is worried about.

THE CENTRAL NERVOUS SYSTEM "SYNDROME"
(Neural Tube Defects)

The anomalies of the central nervous system do not form a "syndrome" in the usual sense of the word. Morphologically and genetically distinct variants are undoubtedly lumped together here, but some of the defects of the central nervous system are clearly related to each other as a result of their dependence upon some genes which they share in common. The commonest CNS malformations are anencephaly, spina bifida, and hydrocephaly. Congenital cases of hydrocephaly may be associated with spina bifida, and spina bifida may be associated with anencephaly. This striking relationship was shown by Record and McKeown [1949, 1950] and is reproduced in Table 16-2.

It should not be forgotten that this "syndrome" is not a homogeneous group and that hydrocephalus without spina bifida may be completely unrelated to hydrocephalus with spina bifida. An example of this distinction is Edwards' [1961] discovery of a few families in which hydrocephalus due to stenosis of the Aqueduct of Sylvius behaved as a strictly X-linked trait. One family pedigree displayed 15 affected males. There is a similar family with nine affected males, clearly due to an X-linked gene, among the records of the Dight Institute. There are also statistical complications depending upon whether the Arnold-Chiari malformation, an occipital meningocoele, or an encephalocoele is present. Most studies did not make these important distinctions so we cannot consider them separately here. As with the cancers, the rare Mendelian type syndromes with clear-cut genetics will sometimes be included along with the large mass of multigenic nervous system malformations (see also Holmes et al [1976] and Carter et al [1976]).

There is a pronounced geographical variation in the incidence of the CNS malformations. The rate for anencephaly is nearly 20 times higher in Belfast than in Paris. While the incidence fluctuates widely between neighboring geographic areas such as Scotland and England, there are also racial differences in incidence with much lower frequencies of all the major CNS anomalies in the Japanese compared with north Europeans, for instance.

We find the usual difference in expression of the syndrome due to sex influence, with about 40 affected males to 60 affected females. This is not a striking sex difference for a multigenic trait and is much less aberrant than the large excess of females with congenital hip disease. Consequently, we would not expect a much higher frequency of affected siblings of male patients (the lesser affected sex) than of female patients. A study by Carter and Evans [1973] had this possibility in mind. Their study of the CNS malformations in London, which has a low incidence of the anomalies, showed the usual marked effect of social class, the neural tube disorders being more common in the lower income groups. Carter and Evans

Table 16-2. The Large Excess of Combinations of the Three Defects of the Nervous System Observed Compared With the Expectations for These Combinations if Due to Coincidence

	Number of combinations expected	Number of combinations observed
Anencephaly with spina bifida	0.98	34
Hydrocephaly with spina bifida	0.75	129

[1973] showed that there was a slight excess of affected siblings of male probands. They found 12/316, or 3.80%, affected siblings of the lesser affected sex (male probands) compared with 13/414, or 3.14%, affected siblings of female probands. They found 4/199, or 2%, of the half-siblings to be affected. Only 0.43% of the first cousins were affected. Thus, for the London study, about 3% of the siblings (first degree relatives) were affected, about 2% of the half-siblings (second degree relatives), and about 0.5% of the cousins (third degree relatives) were affected. These data do not show the sharp drop in affected between first and second degree relatives usually shown for multigenic traits. However, the second degree relatives in this case were half-siblings and perhaps shared a greater within-family environment variance with the proband and his ordinary siblings than would be the case for the children of the probands, who are first degree relatives.

Lumbosacral spina bifida occulta can be detected by x-ray even though there is no external sign of the anomaly. Gardner et al [1974] found that 35.5% of the fathers of children with neural tube defects had spina bifida occulta compared with 25.0% of the control fathers, while 26.1% of the mothers were affected compared with 13.2% of the control mothers. The diagnosis of lumbosacral spina bifida is somewhat subjective, but the above results are in keeping with what one might expect for a multigenic trait.

One of the major studies of the CNS malformations is that of Czeizel and Révész [1970] in Hungary; congenital malformations have become one of the 10 most frequent causes of death, and in 1967 they caused 20 times more infant deaths than infectious diseases. In Budapest the incidence of CNS malformations between 1963 and 1967 was 0.37%, which is fairly high when compared with other countries. They presented a literature search for twin data and using relatively unbiased series came up with 166 sets, both identical and fraternal, with only two concordant pairs. This extremely low concordance could indicate a low heritability or a high rate of early spontaneous abortion of one or both affected twins in many cases. The early abortion of both twins with CNS defects is a reasonable hypothesis when one reflects on the studies of Record and McKeown

[1950] and Böök and Rayner [1950], which showed a substantial excess of still-births and spontaneous abortions in the sibships containing their probands when compared with the control data (see also Janerich and Piper [1978]).

Further evidence that the paucity of concordant twins is probably due to the excess of spontaneous abortions comes from the heritability estimates. Czeizel and Révész [1970] estimated the heritability at over 80% for anencephaly and about 70% for spina bifida. Carter and Evans [1973] provided heritability estimates of about 60%, 65%, and 70% for the South Wales, Glasgow, and London samples, respectively. These heritability estimates are similar to those for other multifactorial traits.

Another test of a multigenic hypothesis is provided by looking at the percentage of affected children born subsequent to the advent of two affected children in a sibship. This percentage should be about twice as high as is the case for children born subsequent to the first affected child, that is, the proband. Fraser-Roberts [1962] and, later, Carter and Fraser-Roberts [1967] carried out the heroic task of finding the risk of recurrence of anencephaly or spina bifida for families which already had two affected children. In the first study there were 53 such children of whom 8, or 15%, had some CNS malformation. In the second study there were 69 children produced after the second child with spina bifida or anencephalus. Of the 69 children 8, or 11.6%, had a CNS malformation. Thus, the risk for a third child with a CNS anomaly is 10%–15%, or twice what is expected if there has been only one affected child in the sibship. This is in accord with expectation on a multigenic hypothesis but not with any other type of genetic mechanism.

The genetic counseling data for the great majority of cases are very simple. For normal parents with one affected child, the risk to a subsequent child is about 5%. For normal parents with two affected children, the risk of a third is about 10%–15%. The risks would be a little higher or lower depending upon the geographic location and racial identity of the proband as well as his social class. Carter and Evans [1973] found that 4 out of 133, or 3%, of the children of adult survivors with spina bifida cystica had neural tube defects. This is an interesting figure because it shows that the children of affected persons have no higher percentage of central nervous system defects than do other first degree relatives, such as the siblings of the probands. Consequently, the risks for first degree relatives of a proband are about 3%–5%, or higher if there are other first degree relatives with a central nervous system defect.

One of the areas of greatest excitement in medical genetics research in recent years has been that of prenatal detection of numerous congenital anomalies. The CNS anomalies are among those for which significant advances have been made. Many of the advances have been dependent upon the utilization of the technique of amniocentesis, which is described elsewhere in this book. The amniotic fluid or the fetal cells within it provide us with the basic materials for study. For neural tube defects the diagnostic indicator is the concentration of alpha-fetoprotein (AFP) within the amniotic fluid. The concentration of AFP rises from the sixth week of gestation to a peak between 12 and 16 weeks and then decreases toward term in normal fetuses. In cases where the fetus has an open neural tube defect the AFP concentration is very much higher than normal and thus is an indicator of the presence of the CNS abnormality. No increase of the *total* amniotic fluid protein is found, but there is a selective increase in AFP. Only open defects, which represent 95% of live-born neural tube abnormalities, can be detected. These defects leak cerebrospinal fluid containing high concentrations of AFP into the amniotic fluid. Neural defects which are very small or are closed or covered by a thick epithelioid membrane may not be detected by this or any other presently available method. Amniocentesis for AFP studies should therefore be done optimally between 14 and 16 weeks' gestation (see the excellent book edited by Myrianthopoulos and Bergsma [1979]).

The neural tube defects are probably more important than erythroblastosis as a cause of genetic burdens to the affected individual, the family, and society. Consequently, the early detection of these defects by the estimation of AFP concentrations in the amniotic fluid is a most significant technical advance. Although the specificity, validity, and applicability of AFP concentrations require further study, it is clear from the extensive literature that the method is an effective and practical way of detecting CNS abnormalities early enough so that the mother of the defective fetus may elect to have an abortion. Even though surgical repair of hydrocephalus (shunting) and spina bifida is available and often reasonably successful, presumably most mothers will wish to abort the defective fetus and try again for a normal child. The decision should involve the father as well as the pregnant mother, and be an informed consent. Presumably the technique would be provided ordinarily only for women at risk, and they should apply for its use as soon as they are aware of their pregnancy.

It is also very important to attempt to eliminate as many false positives as possible. Ishiguro [1973] showed that in 7 of 10 pregnant women carrying twin fetuses, the maternal serum AFP was singificantly higher than the normal range. Therefore, it is imperative that one determine the presence of twins by ultrasonic examination, as described below. Ideally, all samples of amniotic fluid, collected for any reason early in pregnancy, should have AFP estimations carried out. Abnormal estimates would alert the physician to carry out other procedures in an attempt to determine the reason for the deviant estimate of AFP. The review article on AFP during human development and in disease by Adinolfi et al [1975] should be of great help to the physician who has not had previous experience with this technique (see also Brock [1975] , and especially the book by Crandall and Brazier [1978]).

Another useful technique for the detection of the defective embryos is that of ultrasonic diagnosis. Sonography is a method for scanning the uterus and embryo that provides a photograph which must be interpreted by an experienced person. The patient is followed with a series of sonograms from very early in pregnancy.

Ultrasound-guided fetoscopy (see Chapter 6) has a place in the diagnosis of neural tube defects when the results of other investigators are conflicting or inconclusive, and may be useful in assessing the severity of a lesion. Rodeck and Campbell [1978] describe three fetuses where the lesions were clearly seen by fetoscopy in the second trimester.

The easiest technique of all is that of checking the pregnant mother's serum for alpha-fetoprotein concentration. This is an initial screening to select from the mothers at risk those for whom an amniocentesis would be indicated. The best time for detecting open spina bifida by measuring maternal serum AFP is at 16–18 weeks of pregnancy. In the U.K. Collaborative Study [1977], at 16–18 weeks of pregnancy 88% of the cases of anencephaly, 79% of the cases of open spina bifida, and 3% of unaffected singleton pregnancies had AFP levels equal to or greater than 2.5 times the median for unaffected singleton pregnancies. The numbers of unaffected pregnancies with AFP levels above 2.5 times the normal median can be reduced by about a third if women with borderline AFP levels are retested. Women with serum AFP levels above 25 times the normal median at 16–18 weeks will have about a 1 in 10 chance that the fetus has a neural tube defect. The screening is an effective method of selecting pregnant women for amniocentesis for the detection of neural tube defects.

These techniques are of great promise and as they become more widely adopted and more refined they should help prevent the appearance of hundreds or perhaps thousands of CNS abnormalities each year.

The genetic counselor can help increase awareness of these techniques and, while the counselor's function is primarily concerned with the genetic aspects of various traits, he or she can also provide general information needed by the parents of an affected child. There are also other sources of information which the counselor can provide, such as the small pamphlet, "Spina Bifida," put out by the U. S. Department of Health, Education and Welfare. It is published by the U. S. Government Printing Office, Washington, D.C. 20402, and costs 10 cents per copy. It is Public Health Service Publication No. 1023 of the Health Information Service Series No. 103.

Regardless of whether one's bias is toward environmental or genetic factors as etiological agents primarily involved with the CNS anomalies, one can provide effective counseling. All persons who have come to me for information have considered their risks to be much greater than warranted according to any relevant theory of the etiology of the CNS malformations. The risk of a repetition of a

CNS anomaly subsequent to an affected child is about 5%, or about 10% after two affected children. The chance of having a miscarriage or stillbirth at each subsequent pregnancy following an affected child approaches 25%, but this misfortune is less traumatic than the birth of a viable child with a CNS anomaly. Most parents are willing to accept this risk of a miscarriage or stillbirth.

The above data are from the British Isles and Hungary. They provide higher risks than those found by Woolf [1975] in his Utah study. In his material only 0.05% of the births had spina bifida cystica, and only 1.32% of siblings born subsequent to the proband were affected. However, these data permit a heritability estimate of 60%, which is in excellent agreement with those from European sources. Thus, the European risk figures should probably be thought of as maximum values for American populations and Woolf's risk figures as minimal values.

Holmes et al [1976] bring up an interesting criticism of genetic counseling for multifactorial traits. They point out that there is great etiologic heterogeneity of the neural tube defects. The general empiric risk figure for counseling first degree relatives of an affected child is about 3%–5%. However, if diagnosis is competent and cases of Meckel's syndrome, chromosomal anomalies, and other recognizable genetic types of neural tube defects which are recessive are removed from the sample, the risk of a repetition drops to around 1%.

They argue, I think, that the 3%–5% risk figure is therefore useless because most families would have either a higher or lower risk than 3%–5%. They are correct that in the minority of cases the risk may be actually 25% and the 3%–5% risk will be much too low. However, most counselees expect the risk to be higher than 25% and, therefore, the citation of a 3%–5% risk will be helpful to them in decreasing anxiety to a more realistic level. The counselees have no pressing interest in the difference between a risk of 1% compared with 3%–5%.

The neural tube defects usually have a low risk of a repetition but carry a very high burden. The child may never walk, mental retardation may be present, and the apprehension as to the health of future offspring may be extreme. One may wonder how marriages hold together after the birth of a child who survives and has some severe expression of the central nervous system syndrome. A study by Tew et al [1977] of the marital stability following the birth of a child with spina bifida is useful. It was found that the divorce rate for families with a surviving child was nine times higher than that for the local population and three times higher than for families experiencing bereavement of their spina bifida child. It was concluded that a severely handicapped child adds greatly to the strain on a marraige, especially if it is soon after marriage. This strain is diminished by the child's early death.

CIRCULATORY DISEASES

Any multigenic trait runs the risk of being heterogeneous in that it may appear to be a single trait with a fairly specific set of gene loci involved, while in actuality it could be composed of two or more traits with similar phenotypes but with unrelated sets of gene loci as their determinants. One can imagine a trait being composed of two similar traits, one with a prevalence of affected males and the other with a prevalence of affected females. The two opposite deviations from the 1:1 sex ratio might cancel each other, with an approximately equal sex ratio resulting. Heterogeneity will tend to blur or negate the principles of multigenic heredity.

This section will include clearly unrelated circulatory diseases, and it is quite possible that some of them which are considered to be homogeneous traits are actually heterogeneous. The validity of the data in such cases should be diminished to some extent, but the data should still be useful in an empiric risk context. There is a need for better answers. There is also a need for better questions, because we still get requests such as, "What is the genetics of heart disease?" Such a question is too global for any simple answer.

Circulatory diseases are the most frequent cause of death, by far. Coronary and other heart diseases in adults account for the majority of the circulatory defects, but the majority of counseling cases are related to the congenital heart defects.

Congenital heart defects are spoken of as a group of discrete entities, and a genetic risk figure is often given for the group as a whole. Actually, they are a hodgepodge of malformations with but one feature in common — they involve the heart or great vessels. However, some of the individual malformations do appear to be at least partially causally related, as they often occur together in one person or separately in close relatives. Granted all the difficulties resulting from the great heterogeneity of the circulatory diseases, let us see what we can learn about the genetics of the major groups of them, which are of interest from a practical genetic counseling viewpoint.

1. Congenital Heart Defects

It is well known that chromosomal aberrations may result in congenital heart defects. For instance, perhaps only one third of the concepti with the trisomy 21 anomaly (Down syndrome) reach term as liveborns. Undoubtedly, a large percentage of these fetal losses result from heart defects due to the chromosomal imbalance. About half of the Down patients who survived beyond birth have congenital cardiac malformations. Ventricular septal defect and atrioventricular canal are the most common heart defects associated with chromosomal anomalies. Genetic counseling for heart defects that are a component of some chromosomal anomaly would give risk figures which would be lower than those for the chromosomal aberration itself. Therefore, heart defects caused by known chromosomal or environmental factors can be excluded from the heterogeneous group of anomalies which we will consider as a multigenic "trait," knowing of course, that this is possible only as an approximation.

We would like to know what the twin data for congenital heart defects look like because the role of heredity is less obvious for this assortment of anomalies than for a trait like clefts of lip and palate. Nora et al [1967] provided a review of the published data and included material of their own. If we use only the cases for which both the diagnosis of the defect and the diagnosis of zygosity were complete, the data in Table 16-3 remain.

It is clear that the identical and fraternal twins differ significantly, and that there is a relatively important genetic basis for the congenital heart diseases. If we accept a concordance rate of 25% for identical twins and the population incidence of 0.83% found by Mitchell et al [1971], we can read off an estimate from Smith's [1970, p 89] graph of a heritability of about 70%.

Anderson [1977], in a larger and probably less biased sample of twins identified at the University of Minnesota Heart Hospital, found concordance for only 5 out of 61 pairs of identical twins (8.2%), and only 1 out of 45 dizygotic twin pairs (2.2%) were concordant.

The study of Mitchell et al [1971] showed little racial difference in the population incidence of congenital heart defects. They found that the incidence for blacks and whites in the United States was 0.80% and 0.83%, respectively.

Nora and Nora [1978] provided a very interesting table of 15 varieties of congenital heart lesions and the number of siblings of each type, both the normal and the affected. There was a range from 4.2% of the siblings of probands with ventricular septal defects down to 1.0% of the siblings of tricuspid atresia probands who had some congenital heart defect. The average of affected siblings for all 13 lesions was 2.8%. This table, adapted by me, is presented as Table 16-4.

Congenital heart patients survived to become parents in many cases, and Nora et al [1970] found that 210 of their 1,478 probands had produced 571 children, of whom 18 (3.2%) likewise had a heart defect. The children are first degree relatives, as are the siblings who show a similar percentage of congenital heart diseases. The good agreement between the 2.8% of affected sibs and 3.2% affected children of the probands would be expected for multigenic inheritance.

Nora et al [1967] carried out a more extensive study of the relatives of their probands who had atrial septal defects. Most of the affected relatives had the same anomaly. The first degree relatives were found to have 3.5% with congenital heart disease, the second degree relatives had 1.3% with defects, while 0.7% of the third degree relatives were found to have congenital heart anomalies. The drop in frequency of congenital heart defects between the first and second degree relatives is not as great as one might expect on a multigenic theory. However, the likelihood

TABLE 16-3. The Occurrence of Congenital Heart Defects in Twins

	Both twins with heart defects	One twin with heart defects	Totals
Identical	9 (25%)	27	36
Fraternal	2 (4.9%)	39	41

TABLE 16-4. Recurrence Risks When the Proband Has the Following Congenital Heart Defect (after Nora and Nora [1978])

Anomaly	No. of probands	Affected sibs no.	Affected sibs %	Counseling risk (%)
Ventricular septal defect (VSD)	306	28/672	4.2	3
Patent ductus arteriosus (PDA)	220	18/516	3.5	3
Atrial septal defect (ASD)	172	11/380	2.9	2.5
Tetralogy of Fallot (tetralogy)	180	11/366	3.0	2.5
Pulmonary stenosis (PS)	166	10/375	2.7	2
Coarctation of aorta (coarctation)	131	5/281	1.8	2
Aortic stenosis (AS)	155	8/361	2.2	2
Transposition of great arteries	116	4/229	1.7	2
Endocardial cushion defects (ECD)	73	4/151	2.6	2
Endocardial fibroelastosis (EFE)	119	11/286	3.8	4
Tricuspid atresia	52	1/98	1.0	1
Ebstein anomaly	47	1/105	1.0	1
Truncus arteriosus	43	1/86	1.2	1
Pulmonary atresia	36	1/80	1.3	1
Hypoplastic left heart	164	8/370	2.2	2

of dominant inheritance seems small because the penetrance would have to be so low — in the neighborhood of only 5%. Consequently, multigenic inheritance would seem to be the better hypothesis to explain the data.

A somewhat similar study on the various congenital heart defects has been carried out by Anderson [1976]. His data all give lower percentages of affected relatives than those reported above by Nora. Anderson does not speculate about the consistent differences between the two sets of data, which are quite similar in size.

Perhaps one of the most valuable points in Anderson's paper is the caveat that the physician should be most certain that the basis for the heart defect is not a chromosomal anomaly (which would account for about 4% of the cases), a single gene (about 2% of the cases), or purely environmental insults for which there are no firm data as to what percentage of cases result from such accidents.

Only after these discrete causes of the defect have been rejected should one espouse the hypothesis of a multigenic background. However, it is reasonable to expect that the large majority of the cases will indeed be of the multifactorial type, or can be treated as if they are. No doubt there will be some mistakes resulting from such a philosophy, but there should not be many, especially if the above precautions are taken.

The multifactorial hypothesis breaks down when too many genetically different variants are included as one trait. Even though some of the 15 different variants classified as congenital heart defects are probably related to each other genetically, the total heterogeneity is so great that they do not act as a unit trait and fail to meet some of the expectations for a multifactorial trait. Thus, the incidence of a malformation in first degree relatives is expected to be approximately equal to the square root of the incidence of the malformation in the general population. Therefore, if the incidence in the general public is 1%, the square root of it would be 10%. No one has ever observed a total of 10% of congenital heart defects among the first degree relatives of the probands, or even close to 10%. Consequently, it is clear that we are not dealing with a single multifactorial trait. The reasons for the discrepancy are not clear, as Anderson [1976] points out.

Fraser and Hunter [1975] have tackled the problem of the probable genetic relationships between the congenital heart defects. They found an association between malformations in pairs of siblings. An excess of affected pairs was found for 1) tetralogy of Fallot and pulmonic stenosis, 2) tetralogy of Fallot and transposition of the great vessels, and 3) tetralogy of Fallot and ventricular lesions. Their method seems to be capable of revealing etiological relationships, probably genetic, between different types of cardiac lesion.

Anderson [1976] points out that the risk of 2%–3% of a repetition of a congenital cardiac defect in a subsequent child must be increased, perhaps to 5%–10%, once a second case has occurred in a sibship, and perhaps to 10% 20% if a parent is also affected.

It is not my intention to conclude each section with alarms and excursions into unconfirmed experiments, but it is wise to be cautious and to point out that new medical techniques often have side effects such that a trade-off situation develops. The value of female hormones as birth control pills cannot be questioned. The world is a vastly better place to live in because of "the pill." However, Heinonen et al [1977] studied 50,282 pregnancies from the Perinatal Collaborative Project and found that in 1,042 of them the mother had received female hormones during early pregnancy. There were 19 (1.8%) of the subsequent births with cardiovascular defects. Among the 49,240 pregnancies not exposed, there were 385 (0.8%) of the births with cardiovascular malformations. No one would consider elimination of birth control pills even if the above finding were confirmed, but it does point out the need for caution in their use.

2. Rheumatic Heart Disease

The cardiovascular system consists of the heart and pericardium, the blood vessels, and the lymphatics; all are a unit. Embryologically, the heart begins as a thickening in the walls of the dorsal aortae. The heart is fundamentally an enlarged artery and is affected by many of the diseases which afflict the vessels and vice versa. The next condition to receive attention illustrates not only the unity of the circulatory system, but the unity of the whole body. That is, rheumatic fever is a hypersensitivity phenomenon affecting the collagenous tissues and the smaller blood vessels of various organs, with greatest damage usually to the heart, the larger joints, the brain, lungs, and skin. It is a systemic vascular disease which can occur as a sequela of certain types of streptococal infections.

Rheumatic heart disease refers to the acute, subacute, and chronic changes in the heart which either accompany or result from rheumatic fever. It is the next in importance of the circulatory problems *for genetic counseling,* after congenital heart disease. For instance, rheumatic heart disease accounts for about 75% of organic heart disease in patients from 10–19 years of age, whereas it is responsible for only about 5% of heart patients in the 70–79 year age group.

Rheumatic heart disease is unusually interesting from a multifactorial viewpoint because, for once, some of the environmental factors can be identified, at least as far as the streptococcal agent is concerned. The frequency values are highly correlated with geographic areas. The geographic relationship results from combinations of low temperature, high humidity, and high altitude. Cold wet winter and spring months foster rheumatic fever, which has a high incidence in the northern Rocky Mountain states, the Great Lakes areas, and the northern Atlantic area. In addition, the disease appears to correlate highly with low income, overcrowding of sleeping quarters, and any other factors which are conducive to the spread of upper respiratory tract infections.

Group A streptococci are the only organisms that appear to be etiologically associated with rheumatic fever. The concept of an allergic or hypersensitivity reaction to products of Group A streptococci is generally accepted. Presumably, then, we will be concerned with an immune reaction which might be expected to have a "constitutional" or multigenic reaction.

Vesey [1947] showed, with a very large sample of men in a military basic training camp, that 0.21% of the men developed rheumatic heart disease. Of these, 0.14% were white, and only half as many percentage-wise, or 0.07%, were black. Apparently blacks are more resistant to the results of the infection than whites.

Rheumatic-fever-induced heart disease is an excellent example of a disease produced by a genetic-environmental interaction. One could assume that the body, in producing antibodies against the streptococcal infection, may also produce antibodies which could damage the heart. Which people suffer heart damage after the infection has occurred, and to what degree, would depend in part upon their hereditary constitution, that is, their multigenic background.

We have seen that apparently blacks are less susceptible to rheumatic fever than whites. We next look to the twin data for an estimate of the magnitude of the genetic component of the disease. The data from the Danish Twin Register (Hauge et al [1968]) are given in Table 16-5.

The concordance between the identical twins of 33% is fairly high for a multigenic trait and indicates a heritability of better than 80%, using Smith's [1970] graph. A large family study by Wilson and Schweitzer [1954] on rheumatic fever fitted the expectations for simple recessive heredity. However, their conclusions are not acceptable because their sample is so badly biased. For instance, they had eight families in which both parents were affected, and of their 15 children, 14 had rheumatic fever. This is an unacceptable set of data because, as we have just seen, even identical twins are concordant in only one third of the pairs. Fortunately, most of the large literature on rheumatic fever is internally consistent, and it all points toward multigenic inheritance as one might expect for a trait in which the environmental components of temperature, humidity, and overcrowding are so obvious.

In a large study by Stevenson and Cheeseman [1956] 6.4% of the siblings of the probands had rheumatic fever when neither parent was affected, and 11.5% of the siblings of the proband were affected when one parent had rheumatic fever. In a follow-up study of 51 women who had survived rheumatic fever and produced at least one child, there were, again, 6.4% affected offspring. Of the 51 affected women, 7, or 13.7%, had produced at least one affected child. This figure should be of value for counseling persons who have had rheumatic fever and are about to start their reproductive life.

The data fit the multigenic hypothesis in that with 1.0 boys to 2.0 girls affected, there are 14.3% affected siblings when the father had rheumatic fever, compared with only 9.9% affected siblings when it was the mother of the proband who was affected.

It should be mentioned in passing that rheumatic fever and heart disease occur more frequently in non-secretor persons than in secretors, and that serum immunoglobulin A (IgA) concentrations were significantly lower in non-secretors than in secretors (see Grundbacher [1972]). It is not known how the gene which causes secretion of the ABH blood group antigens into the body fluids protects against the rheumatic fever infection, but probably it is related to the IgA concentrations. The relationships, whatever they may be, are relative and not absolute.

In recapitulation of some of the high points of this section, rheumatic heart disease is of great interest because of the interaction of the environmental factors, including the streptococcal agent, with the genetic constitution of the person. The incidence of the disease has been dropping for years as a result of better prophylaxis of upper respiratory tract infections, and one can expect some further improvement from environmental manipulations. The disease is of interest for genetic counseling in that the distinction between the burden to the parents and the empiric risk per child can be made. The data of Stevenson and Cheeseman [1956] described above permit this distinction. They showed that the risk per child of a parent with rheumatic fever was about 6.4%, a value useful for all first degree relatives of a rheumatic fever patient. More important they showed that for any patient who has children, there is a 13.7% chance that each patient will have at least one child with rheumatic fever. This value depends upon the number of children in the family and is only a rough estimate of the true value, but this is of interest because geneticists often fail to present their data so that this dimension of the burden can be seen. It should generally be about twice the risk of an affected child when one parent is affected.

3. Atherosclerosis

The familial occurrence of coronary artery disease has been recognized for one hundred years or more. Atherosclerotic diseases are the major cause of morbidity and mortality in Western society. The ratio of affected males to females is about 3:1. Blacks are much less prone to coronary thromboses than are whites by a factor of about 10 times.

The twin data from Hauge et al [1968] for coronary occlusion are given in Table 16-6.

TABLE 16-5. The Occurrence of Rheumatic Fever in Twins

	Both twins with rheumatic fever	One twin with rheumatic fever	Totals
Identical	60 (33%)	118	178
Fraternal	24 (10%)	214	238

TABLE 16-6. The Occurrence of Coronary Occlusion in Twins

	Both twins affected	One twin affected	Totals
Identical	40 (33%)	82	122
Fraternal	48 (27%)	131	179

The twin data are not helpful in identifying a genetic basis for coronary occlusions, because concordance in the fraternal twins is not only unusually high but also almost as high as that in the identical twins. Both types of twins owe part of their concordance to the high incidence of coronary artery disease in the general population. Presumably, at least 25% of the population eventually dies of coronary artery diseases, so the concordance rate is not much different from the base rate in the general population.

There is another complication which should be mentioned. This is the high frequency of coronary heart disease in patients with hyperbetalipoproteinemia Fredrickson type IIb, which is the commonest familial hyperlipidemia and is synonymous with familial hypercholesterolemic xanthomatosis. This xanthomatosis is considered to be an autosomal dominant trait with high, but not complete penetrance. Slack and Nevin [1968] studied the first degree relatives of 53 patients with xanthomata associated with raised total fasting plasma lipids. The 53 patients with xanthomatosis had 22 cases of ischemic heart disease, or 42% by the time they were studied. They discovered that the relatives of the 53 patients ran such a high risk of early death from ischemic heart disease that early identification and treatment of the children with raised cholesterol levels would be very worthwhile. They found that about 7% of first degree relatives of a proband who died from early ischemic heart disease likewise died of this disease.

The complication described above is that the primary trait is hypercholesterolemia, which is sometimes accompanied by xanthomata and/or coronary artery disease. It is difficult with an assortment as heterogeneous as the heart diseases to determine which anomalies are primary diseases and which are secondary expressions of other traits such as hypercholesterolemia or other syndromes.

Clearly, it would be of great value to establish the diagnosis of hyperlipidemia at the earliest possible age. Goldstein et al [1974] determined the cholesterol and triglyceride levels on 2,000 consecutive umbilical cord blood samples. The mean levels of cholesterol and triglyceride of parents and grandparents of the 126 babies with the highest levels were no different from those of 133 control newborns. They concluded that routine screening for elevated cord blood lipids of newborns does *not* appear to be an efficient method for the early detection of those at risk for development of familial hyperlipidemia and premature coronary disease.

Namboodiri et al [1975] carried out elegant bivariate analyses of cholesterol and triglyceride levels in families in which probands have type IIb lipoprotein phenotype. These did not indicate a general genetic mechanism for both hypercholesterolemia and hypertriglyceridemia, though there was evidence for single gene determinants for each entity. They hope to analyze their data using a refined genetic model that would include both major gene and polygenic effects and to combine this form of analysis with quantitative tissue culture methods (see also Elston et al [1975] and Sing et al [1975]).

There is evidence that low density lipoproteins (LDL) suppress cholesterol synthesis and that binding of the lipoprotein to the specific receptors is a necessary first step toward the suppression (see Goldstein and Brown [1979]). Apparently persons homozygous for the gene for familial hypercholesteremia either have no receptors for LDL or else they have defective receptors that bind LDL very poorly. However, the single-gene form of familial hypercholesteremia afflicts only about one person in 500 and thus is not an important part of the major problem, as pointed out by Marx [1976]. The vast majority of patients with hypercholesteremia presumably have many pairs of genes involved, and there is now some interesting evidence of relationships between cholesterol levels and the ABO, Gm, Hp, and secretor blood groups. These associations are considered by Sing and Orr [1976] to indicate that these marker loci are themselves involved in cholesterol determination or are closely linked to the involved loci. The four sets of blood-group loci are independent of each other and are presumably located on four different chromosome pairs, thus revealing in an elegant fashion a multigenic basis for cholesterol levels.

The elaborate study of Sing and Orr [1978] of the families in the Tecumseh, Michigan, sample estimates that 20% of the observed correlation between full sibs living in the same household is attributable to shared environments and not shared genes. Consideration of common environmental components of variance reduces the estimate of the total genetic component of variability in ln cholesterol from 85% due to additive and dominance effects to 58% due to additive effects only.

Robertson and Cumming [1979] indicate that cigarette smoking is a significant environmental factor affecting serum lipoprotein levels and heart disease. For practical purposes, dietary control of cholesterol levels is the first goal of treatment, and this will be more or less successful depending upon the genotype of the patient. Prevention of cigarette smoking should be a national goal.

Christian and Kang [1977] found heritabilities for plasma cholesterol levels of from 0.52 to 0.68 and also evidence that maternal influences, in part prenatal, are important in setting plasma cholesterol levels.

Our conclusion is that the scope of the problem of the genetics of coronary artery disease is almost too large for our grasp. It is a truly global proposition with great genetic heterogeneity and practically endless environmental factors such as diet, diabetes, personality traits, cigarette smoking, and lack of exercise all contributing to the eventually fatal result. It seems most reasonable to view the problem in the same way we viewed the congenital heart defects, that a few of the types which are relatively rare display simple Mendelian heredity, but most are confusingly multigenic and should be given genetic counseling on a multigenic hypothesis. Coronary heart disease is the major health concern of adults in the United States. While not enough is known about its genetics, it is probable that its origins are during infancy and childhood, and studies of its prevention should start then.

4. Essential Hypertension

This is the last major category of the circulatory diseases. Like the previously considered groups, hypertension used to be thought to result from the the inheritance of Mendelian type genes with lack of penetrance. It is much more reasonable, in view of our present day concepts, to view essential hypertension as a multifactorial trait.

Everyone has a blood pressure because a minimal pressure is necessary for life itself. It follows that the measured diastolic pressures will give a continuous curve which may be skewed because of poor adaptation of individuals at one or both extremes of the curve. One would expect a genetic basis for this universal trait. It would not be profitable to review the polemics as to the type of inheritance involved in essential hypertension because, in agreement with the statement of Hamilton et al [1954], "We suspect that essential hypertension has no real existence as a specific clinical entity. It is a convenient term for those subjects whose arterial pressures exceed a certain level chosen on arbitrary grounds and in whom there is no other disease present that accounts for the high pressure." The hypertensives certainly become clinical problems but the boundary with the normotensive state must be assigned arbitrarily.

One can determine the percentage of first degree relatives of a clinical patient who also are clinical patients, and it will be higher than the percentage of clinical patients who are first degree relatives of a control sample. There are no satisfactory data for such an analysis because there has been no determination of an appropriate cut-off point between normotensive and hypertensive pressures which would have medical significance. Hypertension is not a quasi-continuous trait.

There are, nonetheless, greater risks to the first and second degree relatives of a hypertension proband than to the relatives of a control person. While this fact is not likely to cause any change in the reproductive plans of the person requesting genetic counseling, it should lead to better prophylaxis in the relatives of affected persons. The younger the hypertensive proband presumably the heavier the genetic loading in the family and the greater the need for early and repeated testing of the blood pressure with appropriate medical management when hypertension is detected. The adoption of the genetic counseling "mystique" is merely the application of common sense to a complicated problem.

CLEFTS OF LIP AND PALATE

There is a tremendous mass of statistics which show that in the United States about one out of every 700 live births has a cleft lip, cleft palate, or both. Gorlin et al [1971] and Goodman and Gorlin [1977] list 50 different syndromes, each of which includes some expression of facial clefts, in addition to other anomalies, and which depend upon all types of heredity for their presence in the child. These complicated syndromes are of great clinical interest because of the complexities they provide for the array of health personnel who will deal with them. The great majority of facial clefts fortunately are easier to treat than the syndromes in which multiple anomalies are the primary difficulty, and we will restrict ourselves to relatively simple clefts of lip and palate. Only about 1/10 of cases with cleft lip ± palate have other defects, according to Czeizel and Tusnadi [1971].

Numerous large family studies, beginning with Fogh-Anderson [1942], have shown that close relatives of probands with cleft lip, with or without cleft palate (CL±CP), have a significantly increased incidence of CL±CP but not of isolated cleft palate (CP). Furthermore, the close relatives of a series of probands with isolated CP have a significantly increased incidence of CP, but not of CL±CP. Thus, the genetic basis for clefts of lip with or without clefts of palate seems to be somewhat different and distinct from the genetic basis for isolated cleft palate.

It is reasonable to assume that these variably expressed clefts result from multigenic interactions with numerous environmental factors, though there probably are rare family pedigrees in which a single gene locus is primarily responsible for the anomaly, such as the Van Der Woude syndrome [Cervenka et al, 1967].

We should look at the twin data, as always, to obtain perspective as to the relative influence of environmental and genetic factors for the two categories of clefting (cleft lip with or without cleft palate) and isolated cleft palate. These figures, collected by Gorlin and published in the review by Fraser [1970] are probably somewhat biased toward concordance because of the likelihood that concordant pairs will be of greater interest than discordant pairs.

	CL±CP		Isolated CP	
	Both twins affected	one twin affected	both twins affected	one twin affected
Identical	20 (38%)	33 (62%)	4 (24%)	13 (76%)
Fraternal	7 (8%)	79 (92%)	2 (10%)	18 (90%)

The concordance between the identical twin pairs indicates the importance of environmental factors, while the difference in expression of the trait on the two sides of the body tells us that rather trivial environmental fluctuations within the body of the embryo account for much of the variation. Facial clefts are strictly congenital and are established during the first two months of gestation. Consequently, the relevant environmental fluctuations are all rather intangible and trivial as I demonstrated long ago for clefts in the mouse [Reed, 1936]. This is shown in people by the fact that only about 20% of the cases are bilateral, while of the unilateral clefts almost twice as many are on the left side as on the right side of the face. Even the bilateral cases often fail to be symmetrical, with the cleft being deeper on one side than on the other.

The triviality of the environmental differences is demonstrated also by the fact that although the clefts are established so early in development, almost twice as many males as females have CL±CP. On the other hand, for isolated cleft palate there are almost twice as many female cases as male.

One other significant aspect of facial clefts is that of cleft uvula, which appears to be an incomplete form of cleft palate according to Meskin et al [1966]. However, the incidence of cleft uvula (one per 80 Caucasian individuals) is much higher than that for isolated cleft palate, about 1 per 2,500 births. Like cleft of the soft palate, cleft uvula approaches a 1:1 sex ratio. Cleft uvula is of additional interest because of its high frequency in Orientals and American Indians, while it is much less frequent (1 per 300) in American Blacks, see Schaumann et al [1970]. Cervenka'and Shapiro [1970] showed that 1 in 10 Chippewa school children had cleft uvula. Furthermore, the prevalence increased with increasing proportions of Indian ancestry, but the severity of the cleft in the uvula was not associated with the percentage of Indian ancestry. Their sibling analysis revealed a high frequency (up to 20%) of affected siblings in families with high proportions of Indian ancestry. The racial differences seen here, and for other multigenic traits, are most intriguing. Are the different accumulations of the various gene combinations in the different races related to different strengths of natural selection or are they the fortuitous results of genetic drift? We cannot answer this question yet.

A very comprehensive study of CL±CP and CP has been carried out with the vital statistics of Hawaii by Chung et al [1974], which showed that individuals with Hawaiian ancestry have a higher incidence of clefts than do Caucasians, but no maternal or hybridity effects were detected. They carried out an elaborate segregation analysis, but it provided no clear-cut discrimination between single-locus and multifactorial models for CL±CP or CP. For either condition, the best fitting single-locus model was found to be as good as the multifactorial model in explaining the data. However, the heritability for CL±CP was so high (0.99) that involvement of major genes was suspected for them but not for isolated CP.

Heritabilities for CL±CP have been calculated by Carter [1969], Tanaka et al [1969], Woolf [1971], Czeizel and Tusnady [1972], Bear [1976], and Melnick et al [1977]. The heritability value is an estimate of the percentage of the variability in the expression of the trait, which is due to additive genes and should be about the same value for the various first, second, and third degree relatives. The agreement with expectation for the various relatives and between the English, American, Danish, and Hungarian samples is good, as shown in Table 16-7.

It is important to repeat the caution here that these heritability values are not as threatening as they might seem to be. They are deceptively high, as they include the effects of environmental factors working within the family and those important, but individually trivial, environmental factors affecting the developing embryo. In order to restore our perspective we should recall that the genetically identical twins were concordant in only 38% of the pairs for CL±CP.

TABLE 16-7. Heritability Values by Degree of Relationship

	First degree	Second degree	Third degree
Carter [1969]	0.76	0.76	0.67
Tanaka et al [1969]	0.53	– –	– –
Woolf [1971]	0.76	0.64	0.78
Czeizel and Tusnady [1972]	0.78	0.82	0.84
Bear [1976]	0.60	0.73	1.03
Melnick et al [1977]	0.84	– –	– –

A corollary to quasi-continuous variation is a relation between the severity of the defect and the number of pairs of genes involved. Woolf [1971] demonstrated the corollary with higher frequencies of clefts in the relatives of probands with bilateral clefts than in the relatives of probands with unilateral clefts. Thus, the frequency of CL±CP in the sibs of the probands with unilateral clefts was 3.8%, while the frequency of CL±CP in the sibs of the bilaterally affected probands was 6.7%, almost twice as high.

Family history is of great significance in the risk figures for the siblings of a proband for multifactorial traits. Using Woolf's data again, the risk for sibs born subsequently to the proband was 4.2% if there was no family history, 7.9% if there was a family history other than an affected parent, and 21.4% (n = 14 only) if there was a family history in addition to an affected parent.

Curtis et al [1961] provided similar values for genetic counseling purposes. Where there was an affected parent the risk to siblings of the proband was 16.7% ± 3.3% (in a sample of 126 siblings). They did not find any increased risk in the siblings of a proband whose unaffected parents were consanguineous, the risk being only 3.6 ± 2.5% with a sample of 55 siblings. For a more recent study of such risks see Spence et al [1976].

There are two usual counseling situations. The first one is the question of the risk to the child with an affected parent. There are almost twice as many males in the general population with CL±CP as females. Females, therefore, have a somewhat heavier genetic loading than males, and there would be a higher risk for the children of affected females. Woolf [1971] found this to be true, the affected female probands having 6.8% affected offspring while the male probands had 3.9% affected children. The average risk that an affected person will have an affected child is therefore about 5%. If the affected person has already had an affected child, the risk is increased into the 10–15% range. Melnick et al [1977] do not confirm these findings.

The second and most frequent counseling case is that of a couple who do not have clefts themselves, but do have an affected child. If the affected child is a boy, the risk to the next sibling is about 3.4%, and if the proband is a girl the risk is about 4.2%, according to Woolf [1971]. Both values are in the 3–5% range, which is as precise as necessary for counseling purposes, and the difference in risk according to the sex of the proband is unimportant. However, the effects of severity and family history are substantial and should be taken into account (see above).

A most interesting question is raised by the reproductive fitness of quasi-continuous traits like clefts of lip and palate. In past generations babies with the anomaly died as a result of nursing difficulties, surgery, or other related causes. The cosmetic disadvantage was an impediment to mating and marriage. Koguchi and Tanaka [1976] calculated the relative reproductive fitness for Japanese patients to be 0.22, and for Caucasians the value was 0.17. These low values indicate severe selection against affected persons. One wonders how one can have both severe selection against a trait and such a high frequency of it. The authors have no explanation for this scientific dilemma, either for man or animals of much greater antiquity.

There was once an old Norwegian law which forbade butchers to hang hares in public view for fear that the sight would cause pregnant women to have children with harelip. Fortunately, such old superstitions as "marking" via the optic nerve are gradually disappearing. However, we have new fears that various drugs and radiations, which are our own creations, will cause traits like clefts of lip and palate to appear in embryos that are genetically susceptible to clefts and only need a small exposure to some trauma during the first two months of gestation to exhibit the clefts at birth. The counselor should check on possible environmental insults during the first trimester of pregnancy which might possibly have resulted in the clefts observed in the child. There have been numerous suggestions concerning specific drugs which might cause facial clefts. Safra and Oakley [1975] suggest that diazepam (Valium) taken during the first trimester of pregnancy resulted in 7 out of 15 malformed children having cleft lip with or without cleft palate. This does not give us the risk of a malformation due to diazepam, but it seems to indicate that clefts are among the most likely anomalies if there are any at all.

Little has been said about counseling for isolated cleft palate. One needs only to remember that there is an excess of affected females and that the risks are roughly half those for bilateral clefts of lip and palate.

CLUBFOOT

Our information about the genetics of clubfoot is still in a rather primitive state, partly because of its heterogeneous nature. In the old studies various kinds of talipes were lumped together with the result that no single genetic mechanism would be likely to emerge. The major studies in recent years have concentrated on talipes equinovarus, and the material in this chapter will be restricted to that variety of clubfoot. Cases associated with spina bifida, cerebral palsy, and arthrogryposis are not included.

Clubfoot is the most common serious congenital deformity of the foot. The frequency of talipes equinovarus at birth was found to be about 2.2 per thousand for white males, 1.0 per thousand for white females, 2.1 per thousand for black males, and 2.3 per thousand for black females at birth. These are the values for talipes equinovarus as a single condition and not as part of a multiple abnormality; they are congenital and from carefully studied pregnancies (see Myrianthopoulos and Chung [1974]). Ching et al [1969] found a frequency of 6.8 per thousand for "pure" Hawaiians compared with 1.1 per thousand for Caucasians born in Hawaii. The significant racial effects were contributed by the father's race as much as by that of the mother.

According to Nora and Fraser [1974] there may be malformations associated with talipes equinovarus such as generalized joint laxity (10%), inguinal hernia (17% of affected boys), and minor defects of the extremities (4%–5%).

Few twin studies of clubfoot have been done since the admirable work of Idelberger [1939], which is given below:

	Both members with clubfoot	One member with clubfoot
Identical twins	13 (32%)	27
Fraternal twins	4 (3%)	130

The above concordance rate for identical twins of 32% may not seem very high. However, if there is an important genetic component, Smith [1970] has shown that one should expect about this concordance rate for multifactorial inheritance having an incidence of 1.2 per thousand in the general population, and a heritability of 68% as calculated by Ching et al [1969]. Thus a low concordance rate in identical twins cannot be used to prove that genetic factors are unimportant in the predisposition to a disease. Concordance rates in identical twins will be high, as a result of genetics, only if the heritability is *very* high. Smith [1970] provided a very useful graph for identical twins in which one can compare concordance and heritability for any population incidence of a trait to see if the observations are consistent with each other.

Some human geneticists have asserted that it is not economical to study distant relatives when a multigenic trait is involved. It is expensive to study second and third degree relatives, to be sure, but it is certainly instructive if such information is obtained. For example, the data in Table 16-8 are taken from the work of Wynne-Davies [1965] and from Veale, as published in the book by Carter and Fairbank [1974].

Incidentally, the Maoris in the table above are polynesians and have a general frequency of 6.0 per thousand cases of talipes equinovarus, which is about the same as the 6.8 per thousand in "pure" Hawaiians (see Ching et al [1969]). Orientals, however, have a low frequency of 0.57 cases per thousand (again see Ching et al [1969]).

The major thrust of Table 16-8 is that there is a sharp drop in the frequency of club foot between that of the first degree and the second degree relatives, with a lesser drop between the second degree and third degree relatives. This relative difference in decreasing frequency of a trait in the different degrees of relationship is perhaps the strongest indication that multigenic inheritance is responsible for the trait in contrast to a simple Mendelian mechanism of heredity, where the decreases in frequency should be uniform.

The modern view of the genetics of talipes equinovarus is that of a multigenic etiology with a threshold effect. Therefore, risk figures should increase for subsequent children when there is already more than one affected individual in the family. Palmer [1964] showed that the recurrence risk for siblings where the proband was the only affected person known in the extended kinship was only 1.3%, while if there were one or more affected relatives in the kinship in addition to the proband, the risk was 9.5%. The average risk for the two types of families was 5.0%, a little higher than in most studies.

The usual problem of distinguishing between multifactorial inhertiance on the one hand and dominance with some lack of penetrance on the other is present here, and the more members of a family there are who show the trait the higher the likelihood that a dominant mechanism of heredity is involved. This was demonstrated by the family of Juberg and Touchstone [1974]. Leaving out these rare families with clear-cut simple genetics is important in order to get more valid risk figures.

If we take all the "good" studies into account, we find the following useful risk figures for counseling purposes:

1) If the parents are normal and the patient is a male, the risk to a subsequent sibling is about 2%.

2) If the parents are normal and the patient is a female, the risk to a subsequent sibling is about 5%.

3) If one parent is affected and has one child affected, the risk to a second child is much higher, perhaps 10%–15%.

TABLE 16-8. Talipes Equinovarus: Proportion Affected of First, Second, and Third Degree Relatives

	First degree relatives (brothers and sisters)	Second degree relatives (uncles and aunts)	Third degree relatives (first cousins)
Caucasian (England)	2.9% (8/272)	0.6% (5/823)	0.2% (2/992)
Maoris (New Zealand)	5.6%	2.1%	1.4%

CONGENITAL HIP DISEASE

This category of defects was described by Hippocrates and a hereditary background for it was recognized as early as 1678 by Ambroise Paré. Idelberger [1951] collected data from various sources and found 7,126 boys to 38,485 girls affected, a ratio of 1:5. 4. Record and Edwards [1958] showed incidence rates of 0.2 and 1.1 per 1,000 births, or a 1:5 ratio for affected males to females. Woolf and Turner [1969] reported a sex ratio of congenital hip disease (CHD) of about 1 male to 5 females for the general population of Utah Caucasians. This well established 1:5 sex ratio for CHD makes it a candidate for a multigenic type of inheritance. However, the difference in incidence for the two sexes is by no means proof for a multigenic type of heredity. The obvious differences in the anatomy of the female pelvis, compared with that of the male, might be sufficient to explain the striking deviation of the sex ratio from that of about 1:1 for unaffected persons. For other data and comments see Czeizel et al [1972].

Idelberger's [1951] large study contained a good-sized twin sample which showed a substantial concordance (42%) for the identical twin pairs, which was about 14 times higher than the value for the fraternal twins, as shown in Table 16-9.

Recently, Czeizel et al [1975] reported a sample of twins in which 3 of 6 monozygotic pairs were concordant (50%), while none of the 11 dizygotic pairs were concordant.

Although the concordance between identical twins is 42–50%, the agreement between the two hips of the same person is no greater than that between the one-egg twins. In a subsample of 16,343 cases, Idelberger [1951] found that only 6,558 (40.1%) had both hips affected. The differences in the normal physiological development of the two sides of the body become apparent in the fact that of the 9,785 cases of CHD where only one hip was affected, there were 4,376 (26.8%) with the right hip affected, while 5,409 (33.1%) had the left hip affected. Woolf [1971] showed somewhat greater differences but in the same direction.

All these figures indicate a large discrepancy between the number of persons born with a genetic potential for CHD and those who show the defect. A genotype for the CHD trait is presumably necessary for the trait, but not sufficient to guarantee its expression. Other internal and external environmental factors must be of fundamental significance in the etiology of the CHD phenotype.

What are some of the environmental factors conducive to the expression of the CHD genotype?

Record and Edwards [1958] and Woolf [1971] demonstrated the effect of the stress of breech malposition on the expression of the CHD genotype. While the breech malposition increases the likelihood of CHD, not all breech babies develop CHD. Those that do presumably have a genotype more conducive to CHD than the breech babies who remain healthy.

Woolf [1971] showed that the frequency of CHD in the sibs of the probands born during February–August was higher (4.9%) than in the siblings of probands born during September–January (3.4%). More cases of CHD in the general population are born during September–January than in February–August, so apparently those cases born in fall and winter in Utah have fewer genes disposing to CHD than those born in the spring and summer. Thus, fall-winter seems to be more conducive to the development of CHD than spring and summer. Such seasonal fluctuations are difficult to understand and are of little practical value for counseling purposes.

Record and Edwards [1958] and Woolf et al [1968] observed that CHD cases tend to be first-born children regardless of maternal age, and there is also a tendency for an increased incidence in children born to mothers over 35 years of age. There seems, therefore, to be both a small birth-order effect and a small maternal-age effect for CHD. However, one has to be wary of such findings because they could sometimes result from unknown statistical biases. This is particularly true for CHD, as Carter and Wilkinson [1964] postulated that there are two gene systems, one of which relates to dysplasia of the acetabulum and is multigenic, while the other controls the capsule round the hip joint and may be dominant. Wynne-Davies [1970] found a preponderance of the joint laxity type in neonatal CHD and of the acetabular dysplasia type in late-diagnosed CHD. Czeizel et al [1975] could not show that the neonatal and late-diagnosed cases were two clear-cut entities. The situation is confused.

The study by Beckman et al [1977] showed that in northern Sweden, and apparently in some other areas, the frequency of CHD is increasing and may now exceed 1% of births.

The numerous small environmental effects listed above are probably only a small fraction of those which contribute to the variability of expression of the CHD genotype. The differences in anatomy of the two sexes and the accidents, falls, and other traumas experienced by the baby must be important factors also. We shall see that the heritabilities for males are higher than for females, which indicates that environmental fluctuations are of less importance in males than in females.

One of the consequences of the multigenic theory is that the less frequently affected sex (for CHD, it is the male) should have a higher percentage of affected persons among their close relatives. Woolf [1971] showed that although the differences were not statistically significant, the male propositi did have a higher frequency of affected sibs than the female propositi (5.33% affected sibs vs 3.85%). Czeizel et al [1975] found values of 20.9% affected siblings of male probands and 12.2% for the siblings of the female probands.

CHD is like other traits in that it is genetically heterogeneous. Horton et al [1979] report a kinship in which CHD and increased joint laxity was inherited as an autosomal dominant with nearly complete penetrance. Sixteen males and 16 females in six generations were affected, and there were several examples of male-to-male transmission.

There seem to be racial differences in the frequency of CHD, as is the case with many other multigenic traits. McDermott et al [1960] estimated a prevalence rate of the trait in Navajo Indians of about 1%, which is 10 times higher than the rate in most Caucasians. Is the difference due in part to a racial genetic difference or is it mostly the result of the damage done by the use of the cradleboard by the Navajos? It may not be possible to answer this question as it is now more difficult to obtain support for studies on racial differences than was the case in the olden times, when a typical Navajo family was composed of the father, mother, several children and other relatives, a Harvard anthropologist, and a physican from the Cornell Medical Center!

A more global study on North American Indians is that of Niswander et al [1975]. They examined the records for 43,711 liveborn infants and found 803, or 1.84%, had major malformations. Only 27 cases of dislocation of the hip (0.062%) were listed. This low figure results from the fact that CHD probably would be missed in newborn children under those conditions, as it is not congenital in many cases.

Czeizel et al [1975] state that congenital dislocation of the hip seems to be the most common congenital malformation in Hungary, and they present the astonishingly high frequency of 2.75% in Budapest from 1962 to 1967, and 2.87% in Békés county, Hungary, from 1970 to 1972. They suspect that these rates are high in part because of over-diagnosis. However, it is possible that other studies have resulted in under-diagnosis. Presumably, though, the frequency is high in Hungarians compared with other Caucasians. Furthermore, CHD is also very high among Finns who are also supposed to be of Magyar stock.

There probably was over-diagnosis of a sort in the Hungarian work. They carried out early orthopedic screening, which picked up mild cases that recovered spontaneously in many instances. These were not detected in parents and aunts and uncles, of course, but would be found in the siblings and first cousins of the probands and any others born during the period of study.

The Hungarian group settled on a general population incidence of 1.36% for males and 4.25% for females, and heritabilities of 83% for males and 82% for females. Woolf [1971] calculated heritability values of about 82% for males and 58% for females. There is excellent agreement of the two values for males, while the lower value of 58% for the females is the more reasonable of those two heritabilities, as environmental variables are more important in females.

The usual genetic counseling case will be that of a couple who have produced a child with CHD and are worried about the risk that the next child will have it also. Czeizel et al [1975] presented an elegant table of risk figures for genetic counseling which includes all possible combinations for families of up to and including three children, and with no affected parent, affected mother, affected father, or both affected. These are calculated values and are very high because of the over-diagnosis and the actual high frequency of the trait in Hungary. We can compare the values with those of Woolf [1971] who for U.S. Caucasians, pointed out the increased risks when there is a positive family history.

Czeizel et al show values of about 12% risk for the next child when the affected child is a male, and 9% when the proband was a female with both parents unaffected. Woolf gave comparable values of only 4% and 3%, respectively. If both patents were affected and all three children were affected males, the chance that the next child would be affected, if a female, was 73%. The Czeizel et al table is extremely useful, but for U.S. families the risk figures should probably be reduced to about 50% of those shown in their Table IV.

This writer used to recommend that all siblings subsequent to an affected child be X-rayed in order to detect CHD difficulties early. It might be objected that this is not worth the risk of radiation damage. The study of Cox [1964] on CHD patients who had received multiple X-rays for diagnosis and during treatment is instructive. These persons were CHD probands, all of whom were at least 20 years of age and all were women, as the smaller sample of CHD men patients was excluded. The CHD women produced 201 children, of whom 3 (1.5%) had CHD. This is lower than the 3–4% expected, on a multigenic theory. However, another 26 children (12.9%) had other abnormalities, while among the controls (the nephews and nieces of the proband) there were only 5.7% of other abnormalities and no cases of CHD. The difference is statistically highly significant ($\chi^2 = 8.4$, P < 0.004). The higher frequency of other abnormalities among the children of the X-rayed patients is most distressing and could indicate that they received an unnecessarily large amount of irradiation, assuming the results are not due to some bias or random fluctuations. Fortunately, the siblings would only need one X-ray unless they were affected, and in the latter case the amount of irradiation administered could be reduced substantially.

Most of all "cures" are trade-offs in that there are some undesirable effects of the treatments. In the case of CHD, if the results of Cox given above are valid, perhaps the screening of subsequent siblings should be only by thorough orthopedic examination, without x-rays. This is one of the many crucial dilemmas in genetic counseling. The great value of the screening is obvious, the only problem is knowing how it should be done to obtain the optimum result.

Let me cite one more paper in closing this section because it is challenging. Niclasen [1978] showed that there is a conspicuously high incidence of CHD and Perthes disease in the Faroe Islands. Furthermore, both diseases accumulated in the same kinships. The incidence of Perthes disease was 4.1 per 1,000 males and 0.7 per 1,000 females, while for CHD it was 0.7 per 1,000 males and 5.9 per 1,000 females. Thus, this is a very high incidence of both diseases and an excess of affected Perthes males and CHD females.

Among the 1,123 relatives of Perthes probands there were 10 cases of Perthes disease and 9 cases of CHD. Among the 1,942 relatives of CHD probands, there were 11 cases of Perthes and 23 cases of CHD. Among 5,205 relatives of unaffected probands selected as controls there were only 3 cases of Perthes disease and 10 cases of CHD. Presumably the genetic pool of the Faroe Islands is more homogeneous than in most places because of the small total population of about 40,000 persons. However, that does not explain the striking accumulation of both diseases in the same kinships. How are these two disorders of the hip related?

CONVULSIVE SEIZURES

This is a "hush-hush" group of traits or conditions which have suffered great social stigmata but which, in many cases, fortunately have a "cure," that is, good control of the seizures. However, probably only about 8 in every 10 epileptics has his seizures under good enough control to function normally. There are a number of reasons why the therapeutic promise is not fully realized. There is a scarcity of active, qualified neurologists and adequate treatment centers. The social and economic needs of the patient take a prominent place in the therapy, but the family physician usually cannot meet these needs for the patient.

Convulsive seizures probably involve 16 million people in the United States (about 8% of the entire population) at some time in their lives (see Hauser and Kurland [1975]). They are of much greater social significance than such rare simple Mendelian traits as hemophilia or muscular dystrophy. The last two traits are not blessings by any means, but there is at least sympathy for these patients on the part of the general public which may be lacking for the seizure patient. Fortunately, many of the persons with seizures have only one seizure in their lifetime.

Annegers et al [1976] demonstrated two sex differences for epilepsy which are of the type observed for multigenic traits. There is an excess of affected males. Secondly, the offspring of women with epilepsy have a significantly higher incidence of epilepsy and febrile convulsions than the general population, whereas the offspring of affected men do not.

There is no consistent definition of the term epilepsy, so it will not be used here. The broader term, seizures, will be employed by necessity. Diagnosis is the most important initial step, as with all diseases and disorders. Needless to say, the diagnosis should be made by the top specialist available. Hopefully, there will be a seizure clinic to which the patient can go for initial diagnosis and treatment, at least. The complexity of seizures, diagnostically, is so great that some division of them into groups is necessary before any study of the genetic components can be made. We can start with the group of febrile convulsions, as they are of substantial pediatric significance. They are much more frequent in infants and children than grand mal or psychomotor seizures.

1. Febrile Convulsions

The symptoms, convulsions with fever, appear when susceptible children have a febrile illness. The most susceptible age is between 9 and 20 months, and convulsions with fever are rare before 3 months and after 5 years. The fever is usually high ($>102°$F), the convulsion is usually brief and generalized, but severe convulsions, usually defined as lasting longer than 30 minutes, as well as unilateral convulsions, occur in about a quarter of the children.

It is interesting from a genetic point of view that the expression of the trait is dependent upon having a fever first. Those who develop seizures without having a fever are automatically placed in some other group of seizures. Unfortunately, little is known about the biochemistry of any kind of seizure, but the electroencephalogram (EEG) is a valuable tool even though its relationship to the type of seizure is still fairly vague. There is a conspicuous interindividual variability of the EEG which is useful for genetic research, but there are changes in the EEG during childhood and youth, so the mature pattern is not reached until 19–21 years.

Vogel [1970] has studied the genetic basis of the normal EEG very carefully. He found that, while the identical twins showed no important differences in the EEGs for the two members of the twin pair, there were substantial differences between the EEGs of most fraternal twins. He concluded that even in the higher age groups the EEG is predominantly genetically determined. This valuable tool has not been utilized to the full extent of its potentialities, especially with the febrile convulsions. Hauser and Kurland [1975] reported that 35% of children with febrile convulsions had later seizures, so this trait is not restricted to episodes of fever or to children. They estimated that 3.2% of the children in the community have one or more febrile seizures in their first 5 years.

T. Tsuboi [1977] studied a sample of 6,706 Japanese children and found the incidence of febrile seizures to be 7.2% in males and 6.2% in females, 3 years of age or less. The incidence among siblings was (to me) suprisingly high: 21.9%, or 29.7% when age-corrected. The percentage rose greatly with increasing numbers of affected family members and the segregation ratio among siblings was higher (36.5%) with one FC parent, and lower (18.5%) if neither parent had had a seizure. He also points out the more severe the illness in the proband, the higher the incidence of seizures among the siblings. These are indications of multigenic inheritance or, stated in another way, genetic heterogeneity.

The above data indicate very clearly that seizures during childhood, both febrile and non-febrile, are a very important problem, which, fortunately, seems to be somewhat self-limiting as the result of stabilization with increasing age and with proper medical and social management.

The twin data are of interest in estimating the relative importance of the genetic component in febrile convulsions. Lennox-Buchthal [1973] showed 45%

concordance for febrile convulsions in identical twins. The rate for fraternal twins was not given, but was presumably about the same as in the first degree relatives of the fraternal twins, which was 14%. These values are indicative of a high heritability (about 80%) but do not discriminate between multifactorial inheritance or dominant inheritance of a gene with substantial lack of penetrance. The consequences of genetic counseling are the same in either case.

The second degree relatives included 5% with childhood seizures, while the third degree relatives (cousins) showed 2% with childhood seizures. The drop from 14% affected in the first degree relatives could fit expectations for either a dominant with considerable lack of penetrance or a multigenic mechanism of heredity. Similar results were obtained by Van den Berg [1974] and Frantzen et al [1970].

In concluding this section high temperatures may cause one or more convulsions in those genetically predisposed to this triggering action. Perhaps 3% of the children with one or more febrile convulsions later have chronic recurrent seizures. Presumably fever is only one of the triggering agents to which the genetically predisposed person could respond with seizures. It is probably not necessary to point out that acquired brain damage could result in seizures without there being any increased risk for relatives of the patient.

2. Three-Per-Second Spike and Wave Phenomenon

The febrile convulsive group of patients is selected on the criterion that their seizures were precipitated by high temperature of the body and without particular attention to the EEG. These are almost always patients under 12 years of age. The next group of patients, the so-called "centrencephalic" epilepsy group, are mostly under 20 years of age and were selected on the basis of a specific EEG, which is so characteristic that there is almost universal agreement among electroencephalographers as to its presence or absence from a particular record [Metrakos and Metrakos, 1961]. This EEG is a paroxysmal, bilaterally synchronous, three-per-second spike and wave pattern. There is a partial contradiction in the literature in that a person's EEG is supposed to remain somewhat constant throughout adult life, but the three-per-second spike and wave pattern gradually disappears and is seldom found after age 40.

Metrakos and Metrakos [1961] found that 37% of the siblings of the centrencephalic patients had the three-per-second spike and wave pattern, though only 12.7% of the siblings had seizures. The question then arises as to whether one is interested in the genetics of the EEG pattern or the seizures which sometimes result when the EEG pattern is present. The EEG pattern would seem to depend upon a specific dominant gene with some lack of penetrance, while the appearance of the seizures would depend upon other environmental and genetic factors and therefore should be classified as a multifactorial and multigenic trait.

The aunts and uncles of the proband (second degree relatives) had seizures in 4.2% of the cases, while only 1.5% of the third degree relatives (cousins) had seizures. The control population had only 0.7% seizures for second and third degree relatives, but 2% seizures for the first degree relatives. The latter higher value presumably represents better ascertainment than for the second and third degree relatives of the control persons. The 2% risk should be the proper value for the frequency of affected persons in the general population.

In summary of this section, we find almost identical risk figures for the three-per-second spike and wave pattern probands' relatives as for the relatives of the febrile convulsive patients. Either set of risk figures can be used for genetic counseling purposes. The values of 12%–14% risks for the first degree relatives of the proband are somewhat higher than those for traits such as clefts of lip and palate and other multigenic anomalies. This could result from the monogenic dominant inheritance of a basic defect in some brain mechanisms, which is reflected by the abnormal EEG. The seizures then result if other genetic and environmental factors exert their influence. We have the interesting possibility that we begin with a basic monogenic heredity, which becomes multigenic when the trait is expressed as seizures. However, Gerken and Doose [1973] do not think that the spike and wave pattern is based on an irregular dominant gene inheritance, but instead is multigenic. If they are correct, the process is multigenic (constitutional) from the beginning. A further quibble is that the term "centrencephalic" may not have much acceptance at the present time (see Niedermeyer [1972]).

3. Photogenic Seizures

A photoconvulsive reaction is characterized by the occurrence of irregular spikes and waves in the EEG during stimulation and intermittent light. Such stimulation can result in seizures in genetically susceptible persons. The details of how this comes about are not understood. The frequency of abnormal EEG responses to photic stimulation varies with age and sex, being more common in females and in the 11–15 years age group.

Photogenic cerebral EEG abnormalities occurred in 6.8% ± 0.7% of 1,256 persons who were not epileptic, according to Watson and Marcus [1962]. They found 11.6% ± 1.7% of 372 seizure patients to respond to the photic stimulation. But the percentage of respondents varied extremely – from less than 1% in patients under 6 years of age to 51.7% ± 9.3% in patients 11–15 years old, as opposed to 15.9% ± 4.6% for the non-seizure group (P = 0.0004). When seizures were inadvertently caused by the photic stimulation, they were of the "centrencephalic type."

As Doose and Gerken [1973] point out, "The relatively high incidence of EEG response in the control group, the age dependent penetrance in addition to the preference of the female sex and of the siblings of the female probands indicate a multifactorial system."

It has been known for a long time that some people are likely to respond to photic stimulation with seizures and that some strains of laboratory animals are highly susceptible to seizures from auditory stimulation. Fortunately, these environmental stimuli can be reduced or eliminated without great loss to the person who is found to react to them with seizures.

4. Seizures of Unknown Etiology

The majority of seizures have no identifiable cause. The need to identify a cause of the seizure is particularly important for persons with their first seizure in order to rule out brain lesions, tumors, or whatever causes might be treated so that further seizures could be avoided. Even if meningitis, head trauma, stroke, or a neonatal disturbance such as anoxia antedates the onset of epilepsy, the causal relationship is still speculative.

In spite of their review of all available medical records and a mean follow-up time exceeding 10 years, causes of epilepsy were found in only 23.3% of the 516 cases studied by Hauser and Kurland [1975]. A great deal more clinical research is needed in order to identify the basic causes of seizures.

5. The Trade-Off

There is no actual cure for seizures, but they can be controlled or alleviated to some degree in most cases. Numerous drugs have been used with tremendous benefit by untold thousands of persons. One of the most widely used of these drugs is diphenylhydantoin (dilantin), also called phenytoin. Recently, data have been presented that suggest that dilantin and other anticonvulsants, when taken early in pregnancy, may seriously damage the fetus. This damage has been considered specific enough to warrant the status of a syndrome. It has been named the "fetal hydantoin syndrome" by Hanson and Smith [1975]. Hanson et al [1976] showed that 11% of 35 infants exposed prenatally to dilantin, perhaps in conjunction with phenobarbital or other anticonvulsants, could be classified as having the hydantoin syndrome. An additional 31% displayed features compatible with it.

Among the symptoms observed in some of the children are those such as simian creases on the hands, cardiac anomalies, and clefts of lip and palate which are often found as evidence of any general disturbance of embryonic development. Also present may be hypertelorism, depressed nasal bridge and prominent epicanthal folds, nail hypoplasia, increased body hair, short neck, hair over forehead, and mental retardation. See Anderson [1976] for possible cardiac specificity in the "fetal hydantoin syndrome."

Shapiro et al [1976] looked at the same data as Hanson et al, and they concluded that it may be epilepsy in either parent, rather than the medications, that is related to an increased risk of having a malformed child. The resolution of this difference of opinion is awaited with great interest.

The ethical question is whether anticonvulsants should be discontinued immediately upon determination of pregnancy, or even of expectation of pregnancy. Of course, seizures during pregnancy would be an added danger. Another basic question is that of how many children would a woman on anticonvulsants wish to have? No general answer is useful for these questions other than that as little anticonvulsant medication as possible should be used, and tests to determine the appropriate minimal dosage should be given. An individual judgment will have to be made in each case by the patient contemplating pregnancy and her neurologist. This is an important problem because of the widespread use of dilantin. A less frequently prescribed anticonvulsant, trimethadione, has also been indicated as the cause of fetal syndrome by Zackai et al [1975].

The evidence is even more threatening in the abstract by Dansky et al [1975], who reported that 14 of 88 viable offspring (15.9%) of mothers on anticonvulsant medication were born with the following malformations: 6 children with congenital heart disease, 3 cleft lip, 2 ventral hernia, 1 hypospadius, 1 tracheo-esophageal fistula, and 1 microcephaly. Epileptic mothers not on medication produced 3 of 46 offspring (6.5%) with 1 congenital heart disease, 1 Sturge Weber syndrome, and 1 congenital dislocation of the hip. The monthly anticonvulsant drug levels were recorded, but no significant differences were found in mean anticonvulsant levels between mothers of normal or malformed offspring.

In summary of the material on convulsive seizures, it should have become clear that there is no satisfactory classification of seizures based upon their etiology. The etiology of seizures is so heterogeneous that it will be difficult to provide such a logical taxonomy that everyone will adopt it promptly. Perhaps as many as 75% of seizures would be of "unknown etiology." The genetic situation is not satisfactory, but it is possible to do useful empiric genetic counseling. The first step is to ascertain that the seizures are not a part of some chromosomal or Mendelian type syndrome where the risks may be high for first degree relatives of the proband.

Once the specific seizure syndromes have been eliminated from consideration, we are left with the recurrent seizures which we will assume are behaving like multigenic traits with a rather high frequency of perhaps 2% in the population. It was found that the first degree relatives of probands with febrile convulsions, centrencephalic, or other seizures averaged a 10%–15% risk of having some kind of generalized seizures, though the risk decreases with age. Second degree relatives had a 4%–5% risk of seizures, while the third degree relatives had a risk which was indistinguishable from the population frequency.

There is evidence that mothers who take anticonvulsants during the first trimester of pregnancy have an increased risk that the child will have anomalies caused by the drug. This information should be taken seriously but should not induce panic. A physician who failed to warn young women of this danger, when prescribing any anticonvulsant, might be vulnerable to a malpractice suit. Ignorance of the above information might not be a sufficient defense in the future.

Some idea of the vast amount of literature on seizures, which has been published in the last 25 years, can be obtained from the U. S. government bibliography on epilepsy with key-word and author indexes. This listing contains 1,860 pages and was edited by J. K. Penry [1976].

DIABETES MELLITUS

James Neel et al [1965] stated that, "Diabetes is a geneticists' nightmare." It is perhaps less of a nightmare than some of the other multifactorial traits considered in this chapter, as it has been studied extensively by competent geneticists — which is not the case for all of the multifactorial traits. The nature of the basic defect is still unknown, and the frequency of the disorder in the population is not well defined. One of the statistical problems relates to whether one accepts as diabetics only those persons who are recognized clinical diabetics or whether those are included who have lowered glucose tolerance according to some set of criteria established as a result of tests.

The general public assumes that the discovery of insulin erased the chances of death from diabetes. This is far from the truth. Diabetes still ranks among the 10 most frequent causes of death and is an important disease problem among juveniles and especially for the aged. Diabetes now affects about 5% of the population.

Pincus and White [1933 and 1934] hold priority for the attempt to analyze family pedigrees, with corrections for both the biases due to small family size and to age of onset. After these corrections were made, there was excellent agreement with their hypothesis of a homozygous recessive gene as the basis for diabetes. The advantages of hindsight today suggest that, rather than a single recessive gene in the homozygous condition being the usual situation, we should expect a multigenic basis for the trait. This is rather likely because carbohydrate metabolism is so important to survival that one would expect the genetic background for it to be rather elaborately adaptive to environmental fluctuations. Presumably this would be a multigenic system even if one gene locus were of greater importance than any one of the others.

There is the added complexity that there are probably single gene types of diabetes which are rare and restricted to a few kinships. A family of this sort, in which there seemed to be an autosomal dominant defect called familial hyperproinsulinemia, wes described by Gabbay et al [1976]. Tattersall and Fajans [1975] delineated a group of maturity-onset type of juvenile diabetics whose pedigrees were compatible with autosomal dominant inheritance, although the authors do not exclude multi-

factorial inheritance. There are other such examples of the heterogeneity of diabetes, a concept emphasized by Rimoin [1971] and Rotter et al [1978].

One of the lines of evidence for heterogeneity of the genetic basis for diabetes
cited by Rimoin was the ethnic variability in the frequency of diabetes which can
hardly be explained by dietary factors alone. An interesting study by Doeblin et
al [1969] of Seneca Indians of western New York showed that 11.6% ± 3.4% of
the men and 14.6% ± 3.3% of the women had been previously diagnosed as being
diabetic. No juvenile diabetecs were encountered in the survey. The prevalence of
diabetes in North American Indians seems to be about seven times as great as in
their Caucasian neighbors.

A most interesting paper related to ethnic differences is that of Zavala et al
[1979] who obtained their diabetic probands in Mexico City. Presumably most of
the patients at that hospital were genetically North American Indians or various
racial mixtures. At any rate, the prevalence of diabetes in the seventh decade of
life was more than 13%, and studies of mortality rates showed diabetes to be the
fifth most common cause of death in the 65–74 age group. These figures are similar to those just cited for the Seneca Indians, except that no juvenile diabetics
were encountered in the sample of Seneca Indians. A higher recurrence risk was
found in the Mexican population in the group with an affected mother compared with the group having an affected father.

Goodman et al [1974] also suggest that ethnic genetic heterogeneity exists for
diabetes when comparing the frequency of 35.7/1,000 in Japanese with 12.9/1,000
in Caucasians all living in Hawaii.

Before we consider the genetic mechanism in detail, we should look at the twin
data to see how strong the genetic basis may be.

It is necessary to have an age correction of some kind for twins with a trait like
diabetes. Harvald and Hauge [1965] showed that if one included only twin pairs of
70 years or older, 73% of the identical twins were concordant while 34% of the
fraternal pairs were concordant. If glucose tolerance test results are included, concordance becomes practically perfect in the elderly identical twins. Such high concordance rates are somewhat incongruous for multigenic traits and presumably
either overemphasize the importance of multigenic inheritance or underestimate the
frequency of major genes for the trait in the general population.

Tattersal and Pyke [1972] showed that for identical twins, when diabetes developed in the index twin before the age of 40, half were concordant, whereas in identical twins in whom diabetes developed after the age of 40, almost all pairs were concordant. The unaffected identical twins demonstrated little sign of impaired or
deteriorating carbohydrate tolerance, still less of progression to overt diabetes.
Nor did the unaffected twins show any tendency to produce heavy babies. The
concordant identical twins usually had overt diabetes within three years of each
other. The glucose tolerance test does not seem to be precisely correlated with
the onset of diabetes, as the same person often gives widely flucuating test results over the years before the diabetes is present.

Tattersal and Pike also showed in the above paper that if one parent of a diabetic identical twin is also diabetic, the concordance of the identical twins is substantially higher than if neither parent had diabetes. This is what would be expected if the genetics is multigenic, but would be unexpected for a single gene trait. There would be more genetic "loading," if multigenes were concerned, when a parent is also diabetic, and thus higher concordance of the identical twins.

One conclusion from the twin studies is that if identical twins live to be 85 years old they will almost always be concordant for the presence, or the absence, of diabetes mellitus. One might think that this very high concordance would be evidence for a single gene locus hypothesis but that is not necessarily the case. It does indicate that the genetic influence is very strong because the external environmental influences should become more different as the twins grow older; yet the concordance becomes greater with age rather than decreasing.

Falconer et al [1971] made a thorough study of diabetes in Edinburgh, Scotland. They recorded a prevalence of 0.57% in males and 0.67% in females on January 1, 1968. This included all persons alive there on that day. They found that the morbidity risk increases continuously with age and reaches 2% per annum at the age of 70, after which it levels off but does not decline. This includes only clinical diabetics, as there were no tests for glucose tolerance.

Simpson [1964] proposed a multifactorial mode of inheritance for diabetes and elaborated on the concept in 1969. Smith et al [1972] and Smith [1976] made a careful study of the distinction between the early onset cases (which are usually severe and insulin-dependent) and the late onset cases (which are generally mild and may be insulin-independent). They concluded that, "early- and late-onset diabetes have the same, or a quite similar genetic causation. In other words, they derive from the same distribution of liability to the disease, or have a highly correlated bivariate distribution." The estimates of the heritability of the liability to the disease are higher for the early ages of onset than for the later ages.

Smith et al [1972] also gave low estimates of correlation in liability between spouses which indicated that common environmental effects, after marriage, are not important in affecting the frequency of diabetes in spouses. Their estimates of heritabilities from different groups of relatives were quite variable and were difficult to interpret or account for. There was good agreement for the heritability of siblings and parents, with estimates of around 50%. Estimates for the third degree relatives, that is, first cousins, were around 100%. Such heritability estimates which are 100% or higher indicate single gene action rather than multigenic action.

Goodman and Chung [1975] have reanalyzed the Simpson data and attempted to discriminate between a single locus model and a multifactorial model. They observed, as had Simpson [1969] and Smith et al [1972], decreasing heritability of liability with increasing age. This has been taken as evidence for a multifactorial hypothesis. However, Goodman and Chung obtained a heritability estimate of 99% for early onset diabetes, which is an indication of major gene action. But discrimination between a single locus hypothesis and a multifactorial hypothesis proved more difficult in the middle and late onset groups.

Degnbol and Green [1978] concluded that the risk of siblings or children of early onset diabetics developing the same disease is about ten times that of a normal population chosen for comparison, whereas the risk of siblings developing diabetes later in life does not differ from this normal population. They state, conservatively enough, "that early onset and late onset diabetes cannot have an identical genetic background."

Thus, while diabetes is a clinically heterogeneous group of disorders related to a continuous curve of glucose tolerance levels, and thus a likely candidate for a multigenic model of heredity, the single gene locus model refuses to go away, so we still have something of a geneticists' nightmare and are forced to turn to empiric risk figures for our counseling needs.

Diabetes is a common disease which results in numerous opportunities for marriages between two persons with the disease. Kahn et al [1969] studied the frequency of clinical and chemical diabetes in the offspring of such couples. We would expect 100% of the offspring of such couples to develop diabetes if they lived to be 85 years old and if there were a single recessive gene with full penetrance responsible for all cases of diabetes. This is not a reasonable hypothesis so we expect less than 100% diabetic offspring. The authors cited found only 8.8% of their sample of offspring to have overt diabetes and they predicted that less than 30% would have diabetes by age 85. Glucose tolerance tests of the non-obese offspring showed that about 45% of them had "chemical diabetes," while 62% of the obese offspring had "chemical diabetes." Only about one third of the offspring were over 40 years of age, so the observed frequencies should increase somewhat with age, though it is unlikely that 100% of them would become "chemical diabetics." Similar results were obtained by Tattersall and Fajans [1975b].

Darlow et al [1973] made a special study of the Edinburgh sample of 1,367 diabetics and their 25,635 relatives in order to obtain the empiric risk figures which would be most useful for genetic counseling.

Their data can be summarized as shown in Table 16-10.

TABLE 16-10. Risks (in percentages) to First Degree Relatives in Four Age Categories

	Risks by age of relatives			
	25	45	65	85
Proband onset under 25	8	13	17	25
Proband onset over 25	2	3	9	21

It can be seen that the risks increase with the age of the relative and that they are higher if the proband is under 25 rather than over 25 years of age. All the risks are substantially greater for the first degree relatives of a diabetic proband than for the relatives of a control person. The risks for second and third degree relatives of a diabetic proband are less than half those shown in the table but are still higher than for those for a control sample. These empiric risk figures are well established. Genetic counseling probably will have little effect upon the reproduction of non-diabetics, as ever better control of the disease is to be expected. Therefore, genetic counseling related to non-diabetics will be primarily an informative and educational process with no difficult decisions being necessary as far as genetic consequences are concerned.

It would be nice if we could end the story here. However, there is a "non-genetic" aspect of diabetes which has been ignored rather generally in the literature because of a failure to comprehend its importance from small samples of data. Chung and Myrianthopoulos [1975] used the superb material from the Collaborative Perinatal Project of the National Institutes of Health, which is a prospective type study of 48,437 pregnancies distributed in a dozen areas of the United States. We cannot do justice to their study here, but their data and 339 gravidae from the Joslin Clinic in Boston provided 424 diabetic gravidae on insulin or analog therapy. The 424 products of these pregnancies of white women included 124 (29.2%) with mal-formations of various kinds, which is an astonishingly high rate of malformations. A subdivision by severity gave 75 children (17.7%) with major malformations and the remaining 49 (11.5%) with minor malformations. The data for blacks were less threatening, as only 13.6% of the children of black mothers with overt diabetes had major malformations, while 1.8% of the children had minor malformations only. The major malformations included the usual neural tube defects (such as anencephaly and spina bifida), musculoskeletal defects, cardiovascular, and so on. The minor defects included syndactyly, deformed ears, and many hemangiomas.

In most cases diabetic mothers probably would be willing to ignore the risks of minor malformations, but the 17.7% risk for white diabetic women and the 13.6% risk for the black diabetic women of having a child with a major malformation cannot be ignored. This poses a significant threat to couples where the mother-to-be has overt diabetes.

The risk of having a major malformation in the child of a diabetic woman is in addition to the risk of diabetes in the child later on in life. The major malformations are of the usual multigenic sort, and the increased frequency of them is presumably the environmental consequence of the disturbed metabolism of the diabetic mother. It is reasonable to assume that the long-standing, severe abnormalities in maternal metabolism provided an unfavorable intrauterine environment for the development of the fetus. There is no evidence that the use of insulin increased the frequency of malformations. The nature and mode of action of the unfavorable environmental factors are unknown. The longer the mother had the disease, the higher was the incidence of malformations in the fetus. Paternal diabetes did not contribute an increase in risk of malformations in the fetus.

Generally, diabetes is a disease of adults and is less severe with later ages of onset. If the person has completed his or her family by the time of onset, genetic counseling for this person is rather academic. However, counseling should be of considerable value to juvenile diabetics and other diabetics who still have reproductive aspirations. Zonana and Rimoin [1976] state that, "Unfortunately, accurate genetic counseling for the majority of patients is impossible at present." Accuracy in genetic counseling is always a problem. Even Mendelian traits with complete penetrance provide only probabilities and not certainties that the next child will be affected or not. We cannot claim that no probabilities are available for a trait like diabetes. The client must be given a serious answer with all the qualifications concerning juvenile vs adult disease, ethnic group differences, etc, included. When presented in the appropriate context, the empiric risk figures are the correct, if not completely accurate, material for the client.

One more topic must be considered in relation to genetic counseling for diabetes. This is the relationship of diabetes and the HLA antigen system. The genes in the HLA region on chromosome number 6 have effects which associate them with many diseases, particularly those dependent upon immune responses. It is easy for the non-geneticist to assume that these associations result from genetic linkages, that is, genes for these associated diseases would be on chromosome 6 also. However, this will seldom be the case, as we must expect most of the genes for these associated diseases to be on other chromosomes. Christy et al [1979] show that HLA-Dw2 renders protection against insulin-dependent diabetes mellitus (IDDM), while HLA-Dw3 and -Dw4 are associated with susceptibility. Associations of this sort, even if not genetic linkages, could be helpful in genetic counseling. However, there are exceptions to the associations, which means that the empiric risks will be decreased or increased as determined for the person receiving the counseling. So far, only IDDM has been shown to be associated with HLA. Once again we have a situation where the specialist would be needed to adjust the empiric risks and such experts are not generally available. Once again, though, there is hope for the future of improved genetic counseling.

DUODENAL ULCER

It has been stated that about 10% of all adult males have or have had an ulcer. Chronic peptic ulcer seems to be a disease resulting from urbanization, as it is rare under tribal environments. It is commonly associated with the Jovian business executive who, in his rampaging about, either gets ulcers or gives them to his harried employees.

Peptic ulcers are correlated with hydrochloric acid and pepsin concentrations and are not found in the lower part of the duodenum, which is alkaline. About 40% of chronic peptic ulcers are gastric and 60% are duodenal. However, this ratio varies from geographic site to site and is of no great interest here. We are concerned with whether gastric and duodenal ulcers are genetically independent or not. This problem was resolved to some degree by Doll and Kellock [1951]. They had a large sample of 671 ulcer patients and showed that usually the relatives of duodenal ulcer patients tend to have duodenal ulcers. Only about 10% of the patients had both gastric and duodenal ulcers, but in these cases the relatives tend to have both types of ulcers in the same person, which is not the case for families with only one type of ulcer. Both gastric and duodenal ulcers seem to be multigenic traits, although their genetic components are different. Duodenal ulcers are more frequent than gastric ulcers so we will restrict ourselves to the literature labeled "duodenal" ulcers (DU).

Once a DU has developed it is quite certain that factors such as increased gastric acidity, smoking, and anxiety may make the condition worse, but no one knows why an ulcer arises in the first place.

The figures for incidence of duodenal ulcers vary from place to palce, while the highest incidence of active ulcers in both men and women is between the ages of 45 and 55. In an Edinburgh study of men in this age category, 13.5% were affected compared with 4.7% of the women in this peak age interval. Various studies show a ratio of about 2.5 to 3.0 times as many men affected as women, an indication that we are dealing with a multifactorial trait.

We should look to the twin data, as usual, for a clue as to the importance of genetic and environmental variables. There is an excellent study of Swedish twins by Eberhart [1968] which gave the results shown in Table 16-11.

The concordances are fairly high for a multigenic trait, but they are in an appropriate ratio to each other and indicate a strong genetic basis for peptic ulcers. The values for gastric and duodenal ulcers were not separated, but were stated to be about the same for the two types of peptic ulcers. This research monograph by Eberhart [1968] contains lengthy descriptions of the history of each twin and has a psychological description in each case. It is difficult to estimate the heritability of duodenal ulcers from these data because of uncertainty regarding the population incidence of the trait. If we assume that an appropriate population incidence is from 10%–20%, then the heritability would turn out to be about 75%, using Smith's [1970] graph. Thus, there is a strong genetic basis for duodenal ulcers. The classic paper by Falconer [1965] gave a directly calculated heritability of only 37% ± 6% for peptic ulcer. This was the

TABLE 16-11. The Twin Data for Peptic Ulcers

	Both twins affected	One twin affected	Totals
Identical	17 (50%)	17	34
Fraternal	11 (14%)	67	78

lowest heritability shown for any of the four "traits" he considered. The fact that the two heritabilities are in poor agreement should not concern us unduly, as we must remember that the ulcer data are still rather tentative and that heritabilities are useful only in the context of the specific situation, and are not comparable from trait to trait. They are an attribute only of the population for which they were calculated.

There is no clear-cut dividing line between health and disease for duodenal ulcers, as they are not clearly separable from hyperchlorhydric dyspepsia, duodenitis, or "pseudo-ulcer." Duodenal ulcer is so prevalent in all social and economic classes of western society that the environmental factors responsible probably are operating on the majority of the members of these western populations.

We now come to a remarkable finding which provides insight as to one of the ways in which multigenic heredity works. It had been known for years that patients with peptic ulcer had a remarkably high incidence of blood group type O. The data of Clarke et al [1955] showed that the high group O frequency was related to duodenal ulcer but not gastric ulcer. The association of group O and duodenal ulcer has been found in populations of so many different ethnic origins and in such vast numbers of large samples that it can hardly be due to sampling error. It is quite clear that people of group O are about 35%–40% more liable to duodenal ulcer than are people of groups A, B, and AB. The statistics are unassailable. What is the explanation of this strange association in people around the world?

The first clue came from a study by Clarke et al [1956] on the saliva of 514 patients with duodenal ulcer compared with 491 controls without ulcers. The association of duodenal ulcers with the non-secretor genotype was more marked than that with group O. The data showed that a non-secretor is more than 50% more liable to develop a duodenal ulcer than is a secretor. Combining the results for type O and non-secretor (these are independent gene loci), we find that the people who are O non-secretors are 2.5 times more susceptible to duodenal ulcers than the A, B, and AB secretors.

The mechanisms by which the ABO and secretor genes influence susceptibility to duodenal ulcer are still unknown. The relationships, though interesting, are not helpful for genetic counseling at present. They do demonstrate a significant effect of two independent gene loci (ABO and secretor) on the susceptibility to duodenal ulcer, which is an advance in our understanding of the general problem of multigenic inheritance.

Duodenal ulcer shows a substantial excess of affected males, one of the major characteristics of multigenic inheritance. The next question would be whether the less often affected sex (female) has a higher percentage of affected first degree relatives than do affected males. No study has been designed to provide this answer.

The data are not satisfactory as to the empiric risk figures which would be useful for genetic counseling. However, we do have some rather ancient figures provided by Doll and Burch [1950] which are shown in Table 16-12.

Doll and Burch [1950] showed that familial tendencies were more prominent in early onset cases compared with those age 35 and over. Also, a strong tendency was observed for siblings to have ulcers at the same site as the proband.

In "modern" times, Lam and Ong [1976] classified duodenal ulcer into two subgroups on the basis of age of onset of the disease. Their early-onset group (below twenty years) had a significantly stronger family history, but a frequency of blood group O similar to that of controls, while their late-onset group (over 31 years) infrequently had a family history of ulcer disease, but were more likely than controls to have blood group O. Additional differences between the two groups indicate probable genetic heterogeneity of duodenal ulcers. Consequently, approximate empiric risk figures are the only practical tool we have at present for genetic counseling for duodenal ulcer families. But one must always check to see whether, in the minority of cases, a simple Mendelian mechanism might be involved, as demonstrated by Rotter et al [1978].

The data in Table 16-12 have not been age-corrected, but if they had been corrected for ulcer expectation throughout life, they would have been substantially higher — in all probability. Therefore, we can anticipate for counseling purposes that the first degree relatives of an ulcer patient would have a 10%–20% chance of developing an ulcer during their lives. There is usually a balance between the size of the risk and the burden for multigenic traits, and in this case we have a risk a little above the average, with a burden which is below average. Generalizations of this sort are difficult to quantify, but they are subjectively easy to comprehend. In this case we have a risk perhaps a bit above average for a multigenic trait, but the burden is not high compared with the usual genetic counseling situation. Therefore, while the incidence of peptic ulcer is high, the requests for genetic counseling for this trait will be minimal. Nonetheless, the genetic counselor and the family physician should be aware of the data cited above.

TABLE 16-12. Proved Ulcers in the Sibs of the Probands (site specified in the probands but not in sibs)

Site in proband	No. of living sibs	Ulcers in sibs
Stomach	380	25 (6.6%)
Duodenum	469	36 (7.7%)
Stomach and duodenum	93	9 (9.7%)
Totals	942	70 (7.5%)

PSORIASIS

This anomaly is less of a threat to life than most of the traits considered in this book. Nonetheless, it is a serious disorder and is not merely one of a long list of problems for the dermatologist. The "heartbreak of psoriasis" afflicts between one and three per cent of the population. It has been estimated that psoriasis affects from two to eight million Americans. In persons with psoriasis, skin cells replicate about 10 times faster than the usual 28-day cycle. This excessive growth in some areas of the body produces the red, scaly skin so characteristic of this syndrome. However, the disease is not restricted to the skin, and perhaps the greatest discomfort results from the arthritis which is so often a part of the syndrome.

There has been an active scientific interest in psoriasis during the last few years, both because of the probability of an effective "cure" for it and because of some relationship between psoriasis and various antigens in the blood. The proposed therapy is not yet widely available but, in brief, involves the oral administration of methoxsalen followed by exposure of the patient to 8–10 minutes of high intensity, long wave ultraviolet light from a system especially designed for this purpose. The therapy would have to be under the direction of a physician.

The relationships between psoriasis and various blood antigens are not simple and it is difficult to decide which data are most relevant to an understanding of the relationships. Holzmann et al [1973] suggested that G-6PD deficiency and psoriasis are mutually exclusive as far as geographic distribution is concerned. Ananthakrishnan et al [1974] showed that in the red cell acid phosphatase system there was a significant decrease in the number of heterozygotes in the psoriasis group in comparison with the controls. Beckman et al [1974] confirmed other workers' findings that there is an increased frequency of the HLA antigens Bw17 and B13 in psoriasis patients compared with controls. They also found an association with the MNSs system, where only the Ss factors are involved, and where the data indicate that the Ss heterozygotes may be protected against psoriasis to a high degree. Marcusson et al [1976] concluded from a family study of 26 persons, including 5 with psoriasis, that psoriasis is linked to only one of the HLA haplotypes and that it is not the B17 allele itself which is responsible for the development of disease. Bodmer and Bodmer [1978] show that the most striking association of psoriasis is with HLA-Cw6. There is no good evidence of any immune involvement.

Treatment must be on an individually tailored basis and is life-long. It is said that patients spend more than $1 billion a year on psoriasis remedies. Knowing what not to do in treating the disease is as important as knowing what one should do. Information about psoriasis can be obtained from the National Psoriasis Foundation, Suite 110, 6415 S.W. Canyon Court, Portland, Oregon 97221.

The twin data provided by Watson et al [1972] are as follows:

	Both twins affected	One twin affected	Total
Identical	22 (63%)	13	35
Fraternal	7 (21%)	26	33

Concordance is quite high for a multigenic trait, though one must remember that twin data are likely to be biased toward concordance rather than the lack of it. However, the data are not age corrected. Had this been done, one would expect that the concordance would have been even higher. A glance at Smith's graph [1970, p 89] indicates a heritability of better than 95%. Heritabilities of close to 100% indicate the great importance of major gene effects, and such may be the case for psoriasis.

Steinberg et al [1951 and 1952] suggested that psoriatic patients might be homozygous for two unlinked autosomal recessive genes. More recent studies, such as that of Watson et al [1972], tend to support a multigenic theory. There is no satisfactory way of discriminating between these two suggested mechanisms of heredity; my decision to consider psoriasis a multigenic trait is a matter of judgment. It is not very important for genetic counseling which mechanism is accepted, because we have only the empiric risk figures which can be used for counseling.

Watson et al [1972] made a careful genetic analysis of psoriasis and concluded that the inheritance was mutlifactorial and that the heritability, which was in fair agreement between different degrees of relationship to the proband, was 65%. They think that most of the genetic variation underlying liability to psoriasis is additive. They did not find evidence for any sex differences. They looked at the proportion of the siblings of the probands who were affected when neither parent was affected and found 7.8% affected. If one parent was affected there were 16.4% affected siblings, and if both parents were affected there were 6 out of 12 affected siblings.

The percentages of affected first degree relatives of the probands varied considerably, presumably because of the age of onset effect. The parents included 13.9% affected, the siblings 9.1%, and the children of the probands only 4.5%. The parents had a mean age of 63 years so their value of 13.9% would need very little age correction. Thus we could say that the likelihood of a first degree relative developing psoriasis by age 65 would be about 14%–15%, while for all first degree relatives, without age corrections, there were 209 affected relatives of the probands out of 1,967 persons, or 10.6%. Psoriasis is not a congenital disorder but the large majority of cases will have been diagnosed by age 40. Presumably the empirirc risk figure for first degree relatives most useful for counseling would be 10% rather than the higher value of 14%–15%. The risk figure in the Watson study for second degree relatives was 4.2%, and for third degree relatives it was 1.4%. The last value is hardly distinguishable from the frequency of affected persons in the general population.

It is clear that the circulatory system defects, for instance, are a very hetero-geneous group. Psoriasis is probably a homogeneous disorder as much as any multifactorial disease is likely to be homogeneous. The spread of ages of onset probably results both from the different genotypes expected with multigenic traits and also from environmental differences. The presence of environmental factors is demonstrated by the lack of concordance in almost half of the identical twin pairs. Unfortunately, we have no firm evidence as to what environmental factors are of most importance in the development of psoriasis.

PYLORIC STENOSIS

This is the commonest condition requiring abdominal surgery in early infancy. These cases are treated before the baby is six months old, yet no case is observed immediately after birth. Thus the anomaly is not congenital in the strict sense, but may be so considered for convenience.

Pyloric stenosis is probably the multifactorial "curable" trait for which we have the highest-quality data for study. This is partly because of the characteristics of the trait and partly because the patient who has had a successful Ramstedt operation is at no further physical or reproductive disadvantage. Therefore, the patient later can produce a family which should not be biased toward large or small size because of the trait. The data are also of the highest quality because of the skill and diligence used by Carter and Evans [1969] and Adelstein and Fedrick [1976] in collecting them. As they showed for English Caucasians there was an incidence of about three per thousand total live births, (five per thousand males and one per thousand females).

The excess of affected males does not indicate recessive sex-linked heredity, as the large number of affected sons born to affected fathers rules out X-chromo-some transmission of the trait. It is a case where a trait is sex-influenced; as many females as males have the genetic basis for the anomaly but the differences related to being female protect them from expressing the anomaly in many cases. Fe-males have to have more genes for pyloric stenosis in order to develop the anomaly than do male babies.

Unfortunately the twin data for pyloric stenosis are not satisfactory. They are either selected cases with a bias toward concordance, or in the two probably unbiased samples from MacMahon and McKeown [1955] and Carter and Evans [1969] there seems to be an excess (no doubt coincidental) of discordant twins.

The two unbiased samples are combined below:

	Both twins with pyloric stenosis	One twin with pyloric stenosis
Identical	4 (21%)	15
Fraternal	3 (6%)	51

There is a strong genetic component for pyloric stenosis, not withstanding the low concordance of the identical twins of only 21%. This was demonstrated by the data presented in Chapter 15, and it is not necessary to repeat those figures here except to state that Falconer [1965] calculated a heritability value of 79% ± 5% for pyloric stenosis. We should look at the relationship of the frequencies of the different degrees of relatives, however. The parents of the probands will be omitted, as the Ramstedt operation was not uniformly available for their generation. This is also true for the aunts and uncles. Consequently, our first degree relatives are composed of the children and siblings of the probands, 71/1,647 (4.3%) affected. The second degree relatives are the half siblings and the nephews and nieces of the probands with 13/838 (1.6%), while the third degree relatives are the first cousins and 46 of them were affected out of 8,469 in all, or 0.5% affected. These figures are close to expectations for a multifactorial trait with a frequency of 3 per 1,000 live births in the general population. The data used above are from Carter and Evans [1969].

Carter and Evans [1969] calculated the heritability values for first degree relatives to be 76%, with 65% for the relatives of the male probands and 85% for the relatives of the female probands. The heritability was about 61% for the second degree relatives (nephews and nieces only) and 50% for the third degree relatives (first cousins only).

The table for pyloric stenosis given in Chapter 15 showed very clearly the heavier genetic loading of affected females compared with affected males. This is extremely positive evidence that pyloric stenosis is a multifactorial trait of the quasi-continuous type.

It has been shown in the past that there was an excess of pyloric stenosis among first born children compared with the other birth orders. However, Huguenard et al [1972] demonstrated that these early findings were a statistical artifact. They pointed out that a disease which is more common in an ethnic or socioeconomic group, which has a smaller mean family size than the population as a whole, will tend to show a higher incidence among first born infants when compared to a sample drawn from the whole population. This is the situation with pyloric stenosis. Adelstein and Fedrick [1976] showed that the risk to second and third born infants was higher than that for the first born with the lowest incidence at higher parities.

There is no suggestion from the data that consanguinity is of significance for pyloric stenosis, nor would we expect much evidence of a consanguinity effect for a "curable" trait with a frequency as high as 3 per 1,000 live births.

Little is known about the frequency of pyloric stenosis in other racial groups. However, it is less frequent in blacks and Asians (see Leck [1976]).

There are four significant counseling situations which the physician is most likely to encounter. Using the empiric risk data from Carter and Evans [1969] and the calculations of Bonaiti-Pellié and Smith [1974], we can provide the following risk figures for the four situations.

1) The parents are not affected but have one affected child. The empiric risk that the next child will be affected, if the proband was a male, is about 3.2% (36 affected out of 1,111). This is a low risk and practically all couples will take it, if they had anticipated having one or more additional children after the birth of the proband. The empiric risk, if the proband was a female, is about 6.5% (35 out of 536). The couple would probably also find this risk acceptable unless they had intended to complete their family with the birth of the proband. Because of the success of the Ramstedt operation, the "burden" of pyloric stenosis is not too great; thus couples will be willing to take these small chances of a repetition of the trait in a subsequent child.

2) One of the parents-to-be was affected with pyloric stenosis as a baby but the other was not affected. If the affected parent-to-be is a male, the empiric risk is about 4.0% (27 out of 684) and the calculated risk from the table is the same, while if the affected parent-to-be is a female, the empiric risk is about 13.1% (27 affected out of 206 children) while the calculated table gives a value of only 5.1%.

3) Neither parent was affected as a baby, but they have had two affected children. The risks vary from 8.6% if both affected children were males to 13.0% if both were females. Naturally, there will not be many families of this type, but if there are any, they will certainly ask their physician what the risks are and the doctor should be able to supply the above answer.

4) In this case an affected parent has produced an affected child and wants to know the risk for the next child. If the affected parent is a male and the affected child is a male, the risk is 10.2%, while if the affected parent and child are both females, the risk is 12.8%. The other two combinations of sexes for parent and child will have intermediate risks. These values are from the Bonaiti-Pellié table of risks, while Carter and Evans [1969] state that if the affected parent was a male, 3 out of 17 offspring (17.6%) subsequent to the affected child had pyloric stenosis. If the affected parent was a female, 5 out of 13 (38.5%) of the children born subsequent to the affected child also had pyloric stenosis. Both of these observed values are substantially higher than those from the risk tables and the reason for this is not apparent. The latter high risk might be the result of a maternal effect as suggested by Kidd and Spence [1976]. The maternal effect, if confirmed, would be over and above that resulting from a multigenic contribution.

The last observations cited above would be very discouraging to a family, even though both the observed and expected risks are less than those for a regular dominant Mendelian trait such as Huntington's disease or cerebellar ataxia. However, numerous claims of associations of one sort or another between pyloric stenosis and say, maternal blood groups such as found by Dodge [1974], fail to be confirmed by others, in this case by Adelstein and Fedrick [1976]. Therefore, one should wait for confirmation of the Kidd and Spence [1976] findings before being too discouraging in discussing the situation with the parents. Lalouel et al [1977] contradict the conclusion of Kidd and Spence [1976], and the former authors use Morton's "mixed model" hypothesis to provide support for a multifactorial and multigenic concept of the etiology of pyloric stenosis. As Lalouel et al state, there is much to be learned about the etiology of pyloric stenosis.

17
Normal Traits

What is a normal trait? Any trait such as Huntington disease is clearly an abnormal trait; there is no question but that the dominant gene for Huntington disease is an undesirable gene, and it is difficult to conceive of a situation in which any possible advantages of this gene would outweigh the obvious disadvantages of the gene when expressed in the person involved. It is possible to list a few thousand such abnormal traits, as has been done by McKusick [1978]. The quasi-continuous traits such as clefts of lip and palate, which vary greatly in their severity and depend upon gene differences at more than one locus, are likewise distinguishable from the normal condition. But where does one draw the line for a trait such as height, where we have a normal curve type of distribution from the shortest to the tallest person? Clearly the very short and the very tall are at varying disadvantages, as our facilities are designed for those of average height and not for persons at the two extremes of the range.

How do we decide where normality stops and abnormality begins? The person with an IQ of only 20 may be part of the normal curve of intelligence, but he cannot partake of most cultural activities; he is considered to be abnormal and will be brought to a physician for treatment of his "disease." If no treatment is available, he may spend his life in a state or private hospital. Clearly, the decision as to where normality leaves off and abnormality begins is arbitrary. This arbitrary choice is usually determined on statistical grounds. It is known that if one measures the heights of a group of people, the items will form a bell-shaped curve, which, if it is a normal curve, will be symmetrical. That is, one-half of the people will be found to have average or less than average height, and the other half of the population will be of average height or taller. The person who is in the middle of the distribution of heights is said to be of median height, and this is the same as the average height if the measurements of all the heights form a symmetrical, normal curve.

If one starts at either end of the normal curve, it can be seen that the curve is concave and then at some point it changes its shape and becomes convex. At this point, on both sides of the curve, a perpendicular line may be dropped to the horizontal axis. The number of measurements of heights, IQs, or whatever, between the two perpendicular lines, will include two-thirds of the total population. This two-thirds of the population is between plus and minus one standard deviation from the mean. One can measure off a second standard deviation of the same distance from the first standard deviation, on both sides, and erect perpendiculars to intersect the curve. If this is done, it will be found that 96 percent of the individual measurements of heights, weights, or other traits will be included between these new perpendicular lines — that is, between plus and minus two standard deviations from the mean. Finally, over 99 percent of the population lies between plus and minus three standard deviations from the mean. This leaves less than one percent of the population to be divided between the tails of the curve. Thus, less than one-half of one percent of the population is so tall as to be beyond three standard deviations from the average-sized person.

Our arbitrary answer to the question of where does abnormality begin is that it includes those persons beyond plus and minus three standard deviations from the mean for the trait. There may be exceptional persons in the "abnormal" groups who function very well.

What kind of genetic basis for the normal curve types of traits would we expect to be operating? This is not a new subject for study because "continuous" traits, as distinguished from "discontinuous" traits, have been of great interest since genetics became a science. Continuous traits are, of course, multifactorial. If there is a major gene involved in segregating families, it may be possible to detect the presence of this major gene, as the data should show a bimodal distribution rather than a normal curve. Discriminant functions may be developed to demonstrate the existence of such a major gene, as has been done by Elston et al [1975] for hypercholesterolemia. In a study of height there will be important major genes involved for a small percentage of the population, such as those with ateliotic and achrondroplastic dwarfism; however, the mass of the population will be expected to demonstrate multifactorial inheritance involving many gene loci and many environmental effects. The individual effect of each gene pair or each environmental factor will be small.

The dilemma for the human geneticist working with normal traits is that he must use the statistical methods, such as heritabilities, which are less reliable with human populations than with the well-controlled agricultural strains for which they were developed. Consequently, we must be cautious in our interpretation of whatever findings we might obtain from our data. Our data are often of poor quality and do not merit treatment with highly sophisticated statistical methods and we can easily come up with nonsense answers if our methods are better than our data. If the methods are elegant and the data are inadequate, we can be misled into a loss of proper perspective, as has occurred in the case of some studies of the inheritance of intelligence. Keep an open mind and don't believe everything that you read, even in this book.

HEIGHT AND WEIGHT

Height and weight are not highly correlated. They are, however, very much interrelated. According to Went [1968], "if a man were twice as tall as he is now, his kinetic energy in falling would be so great (32 times more than at normal size) that it would not be safe for him to walk upright. Consequently, man is the tallest creature that could reasonably walk upright on two legs. The larger mammals can become taller, because they are more stable on their four legs. Yet they break bones more easily when they fall."

There is a well-known tendency for mammalian species like the horse to become larger as a result of natural selection over millions of years. Greater size was advantageous, but as can be deduced from the concept in the paragraph above, a limit will be reached beyond which the advantage of greater size will be negated by the greater risk of broken bones from falls. The average height of the species may change rather substantially in a few generations, as it has in the United States, due to the discovery of vitamins and other technological improvements in nutrition. However, the change in the average genotype regarding height will be slow, and natural selection will restrict the genotype within a range determined by the environment.

It is impossible to have height without weight, so there is a basic and inevitable correlation between the two. How large is it? The correlation between height and weight for men is only 0.48, which means that there is a vast number of different combinations of heights and weights, as everyone knows. These different combinations result from the differences in genotypes and environments interacting together.

What kinds of problems result from the multifactorial basis for height and weight for which genetic counseling can be helpful?

The problems associated with height are probably less severe than those concerned with weight. There are numerous varieties of weight-watchers groups but no height-halters organizations. While tallness seems to be favored, shortness is considered a disadvantage, and there are promising therapies for some types of dwarfism. Correct medical diagnosis is the greatest problem for genetic counseling for the Mendelian types of dwarfism. For instance, it is absolutely necessary to determine whether a child has the classical Mendelian dominant type of achondroplasia or the Mendelian recessive type of diastrophic dwarfism. The former case may result from a new mutation in a small sector of a gonad of either one of the unaffected parents, and the risk of a repetition of the trait in a subsequent sibling is probably about one percent. However, if the child has diastrophic dwarfism, which behaves as a recessive trait, the risk of a repetition is 25% for each subsequent sibling, as both parents must be carriers of this recessive trait. The difference between a risk of one percent and 25 percent is substantial, and a failure on the part of the genetic counselor to obtain the correct diagnosis would be unpardonable.

The more frequent problem with small size is not the result of these rare types of dwarfism but the inevitability that with any multigenic trait there will always be some persons at the far end of the normal distribution. When the nutrition of the whole population improves, as it certainly has for the last few generations of Americans, the size of the whole population shifts to the right, and the present generation is taller than the previous one. The relative percentage of persons at both ends of the curve stays about the same, however, and the disadvantage of being in either tail of the distribution remains. The offspring of two very short persons should be taller than their parents, on the average. The offspring are expected to obey Galton's law and regress toward the mean height. However, an occasional offspring is expected to be shorter than either of his short parents.

Galton's law of regression toward the mean also works at the tall end of the normal distribution of heights. The offspring of two very tall parents can expect to have an average height less than the average for their parents. However, an occasional offspring will be taller than either of his tall parents.

The fact that most children of very short or very tall people will be closer to average height than their parents is a very reassuring bit of information that the genetic counselor can provide.

Parents are very much involved with the feeding of their children, regardless of any special height considerations. The "hassle" starts when the child is born. Presumably nursing the child is advantageous to both the mother and the child. However, the usual pressure on the mother to nurse the child seems to be out of all proportion to the expected benefits compared with bottle feeding. Overeating or undereating are psychologic weapons, which even the youngest children use to obtain what seem to be profitable concessions from their parents. If the feeding of the child is considered to be a simple, normal process rather than a continuous bargaining opportunity, a great deal of mental stress will be avoided. Generally speaking, a person's size will be determined by his genotype and the nutritional environment in which he was reared.

What does heredity have to do with body size and shape? The ancient Greeks were interested in the subject and realized that there were different body types and had names for them. Today we classify body types in general as ectomorphic, mesomorphic, and endomorphic. The first extensive family studies of body size and shape were conducted by Davenport [1923]. His work shows that when both parents are slender, the offspring usually are slender. When both parents are fleshy, they have offspring showing a much greater variability of types, from very slender to very fleshy. The genes causing slender body type seem to show recessiveness to those for fleshy body types. The data are adequate to demonstrate at least three pairs of genes that are concerned with the major hereditary variations.

If a person's eating habits are imposed upon him by his mother, who usually selects and prepares the food, the person's body build should resemble that of the mother more than that of the father. If, however, the child selects his food from what is available, and most children are very selective, the body build should be the result of the person's genotype to a large degree and therefore be about intermediate between that of the mother and the father, on the average.

The writer has Davenport's original data at the Dight Institute and has carried out a bit of research on them. Davenport used an index number of weight in pounds divided by stature in inches squared, all multiplied by 1,000. This gave a series of index numbers for individuals, ranging from 20 for the very slender to 60 and over for the very fleshy. The sons and daughters were postadolescent, and adjustments were made for the age differences of the parents and progeny. The results are given in Table 17-1.

The fact that the body build indices of the progeny shown in Table 17A−1 are almost exactly intermediate between those of their parents, regardless of whether the mother was thin or fat, indicates that the most important variable for body build is multigenic inheritance.

The Davenport index of body build was an early kind of somatotype. Somatotyping has proved to be an important, or at least amusing, discipline, and it is appropriate to recommend the small book by Parnell [1958] as an introduction to the subject.

One practical use of a rough kind of somatotyping is known as the Wetzel grid. The pediatrician plots the weight/height index of the child from time to time on the Wetzel grid. If the plotted points stay within the appropriate channel on the grid, then all is well. If the latest point departs from the channel, the physician should search for organic or psychologic problems.

Perhaps 40 million adults and children in the United States are overweight. This means that they are fat. They have more fat than is necessary for their best health. This is about one-fifth of the population and is not so high due to some abnormality, but rather, it results from the mere availability of food. People eat more than they need to. Most of the overweight persons could eat less and lose

TABLE 17-1. A Demonstration That the Body Weight of the Progeny, as Measured by Davenport's Index, Is Intermediate Between That of the Parents. (Indices for 380 pairs of parents and their 1,267 children)

Average index	Mother thinner than father	Mother fleshier than father
Mothers	30.4	40.2
Fathers	38.4	32.8
Parents	34.4	36.5
Daughters	33.7	35.4
Sons	35.6	36.3
Progeny	34.7 ،	36.0

weight, if they so desired. Many do "diet" and keep their weight from becoming a hazard to their longevity. A small percentage of the population becomes, and remains superobese. Severe obesity seems to be a habituation syndrome and it is even more difficult to overcome than alcoholism. Swanson and Dinello [1970] reported on an interesting group of 25 persons who weighed 300 pounds or more who were voluntarily hospitalized for weight reduction. They were given no calories for eight to 85 days (average 38 days); that is, they starved for that length of time. They each lost about 100 pounds before release from the hospital on a 600–1,200 calories per day diet. Most of them were confident that they could continue the 600–1,200-calorie diet, but none did. Within one year after the hospital starvation, all but four were at a weight equal to or exceeding that upon admission. Their rationalization of their failure to continue their diet was that denial of their need for food demanded so much concentration that they had insufficient energy left to meet their other life demands. They were aware that, for them, food was very effective in decreasing their tensions and frustrations, including their annoyance at being obese.

The precise genetic basis for these superobese persons was not studied. However, it is obvious that there is some genetic basis for their habituation to overeating that can be overruled by the environment only with such heroic measures as hospitalization with starvation. They cannot, by themselves, restrict their eating sufficiently to remain at a desirable weight. Most individuals in the population have a feedback mechanism that causes the person to turn off food consumption before too much has been ingested; this turn-off mechanism apparently fails to function in the superobese.

A recent study by Hartz et al [1977] on the relative importance of the effect of family environment and heredity on obesity is instructive, in that it showed about one-third of the variation in the weight of children from obese mothers was due to family environment. Their results suggest that family environment, which consists of such things as parental example and child-rearing techniques, has an important effect on child obesity. Child-rearing practices can be controlled to some degree, but there is no way yet in which the child's genes can be altered.

SKIN COLOR

One can justify the inclusion in this book of a section on skin color by the following quotation from Harrison [1973] : "More is known about the inheritance of skin color than about any other quantitatively varying character in man." If this statement is true, it is certainly an unusual situation because the data upon which it is based were collected by quite early workers, except for his own. Davenport and Danielson [1913] concluded from a pedigree analysis that two pairs of genes were responsible for skin color in the Jamaican families studied. Stern [1970] raised the number of gene loci involved in Negro-American crosses to three or four, using the data from Herskovitz [1930]. Harrison [1973] used data collected much more recently in Brazil and concluded that there were three or four pairs of equal and additive gene loci responsible for black-white skin color differences. He estimated that environmental factors caused about one-third of the variance in skin color and that two-thirds of the variance was due to genetic factors.

My good friend T. Edward Reed [1969] has shown that the Negro-American population probably has an average of about 20 percent Caucasian alleles in its genetics. There is thus a substantial amount of genetic variation in black Americans, particularly in the northern states. What have skin color and racial admixture to do with medical genetics? Perhaps nothing directly, though indirectly there is an important relationship. The physician working with the childless couple is often forced to give up hope that a particular couple will produce a child. If it is the wife who is sterile, artificial insemination (semi-adoption) is of no help. There are not enough "blue-ribbon," newborn Caucasian babies available to even approach the demand for them. There are usually some products of racial crosses available, and these children should be excellent adoption risks. In recent years it has become "fashionable" for white couples to adopt such children. Until the childless couple actually get a child, their needs have not been met, and the physician should be willing to help with their adoption problem. The physician may be asked to evaluate the child in terms of its physical and mental health and also perhaps to predict what the subsequent skin color and general appearance will be. The adoptive parents expect to make a large emotional investment in the child and have a right to whatever scientific information the physician can provide.

The reason that the adoptive parents are concerned about the skin color of the baby is that the child will go to a school in their neighborhood and presumably will marry into the white community. They wish to anticipate and adjust to whatever social problems they think might develop.

Incidentally, there is no truth in the "black baby" myth. Two Caucasians never produce a black baby. The union of a person with mixed racial ancestry and a Caucasian can produce only children intermediate between the parents. The offspring of two mixed racial parents can segregate into various genotypes, ranging from children darker than either parent to children lighter than either parent.

There is little reason to expect any genetic "surprises" from adopted children of mixed racial ancestry. In fact, genetic weaknesses peculiar to one of the ancestral races might be covered up by genetic strengths in the other race. For instance, the genes for phenylketonuria and cystic fibrosis are more frequent in Caucasians than in Africans. The gene for sickle cell anemia is more frequent in Africans than in Caucasians. The hybrids could expect to be free of all these recessive disorders which have different frequencies in the parent races. For details of a follow-up of adopted children of mixed racial ancestry, see Reed and Nordlie [1961] and Scarr-Salapatek and Weinberg [1976].

In general, a child will resemble his parents if they are similar in pigmentation, and be somewhere between them if they differ.

Skin color, because of its visibility, has a social importance completely out of proportion to whatever biological significance it may have. Black skin color implies genetic relationship to those who once were slaves and who are still working their way to full social opportunity and civic responsibility. Other indications of African ancestry, always open to error, of course, are the following:

1) The sacral spot. This is a blue black area at the base of the spine which extends out onto the buttocks. It is present at birth but disappears within a few months or years. It is found in pigmented peoples, including Caucasians with dark complexions. If the spot is absent in the baby, it is evidence that no genes for skin color are present either.

2) Finger smudges. Babies with a small amount of African ancestry often show "smudges" of pigment on the backs of the fingers between the joints. While there may be no major gene for skin color present, presence of the smudges indicates that the child will develop a dark complexion. Absence of the smudges is a good indication that the skin color will remain white.

3) Two-tone babies. In some babies the area around the knees and in general the lower half of the body may be darker than the top half if African heredity is present.

4) Pigmentation of genitalia. If the genitalia are appreciably darker than the surrounding skin, some African ancestry may be indicated.

5) Darkening of skin and eye color. This may occur during the first year of life.

Let me make a final comment about the genetics of the American Negro hybrids. The traits for which blacks and whites can be seen to differ are mostly of the multigenic types — that is, skin color, hair shape and texture, nose width, and so on. Consequently, there is lack of dominance for all of them because, with multigenic inheritance, the relevant alleles are additive in nature and do not display dominance. Thus each differing allele shows its own characteristic in the hybrid person. This is why the offspring of a racial cross will be roughly intermediate between the parents when one looks at the composite of all the differing traits.

INTELLIGENCE

The physician is well aware of the ancient arguments about the nature-nurture problem, particularly as it relates to intelligence. The perplexing question of how much of a child's disease process is the result of his genetics and how much due to his nurture is certainly of interest to the pediatrician and the family physician. The answer to the nature-nurture question remains elusive. It is hoped that this brief consideration of the very broad topic of intelligence will provide a better understanding of the role of the parents as providers of both the child's environment and heredity, not only for intelligence but for all the other traits where the nature-nurture question arises.

The nature-nurture question has substantial political and social components. An interesting paper by Halperin et al [1975] demonstrates one small aspect of this dimension in relation to intelligence. Halperin collected data from Russian twins in 1936 in a society that deprecated class structure. Therefore, family environment including social status, might have had a less important effect on intelligence than in contemporary American society. However, comparison showed that the effect of family environment on intelligence was not significantly less important in prewar Russia than in contemporary America. Russian and American intelligence seem to be no different in spite of the political differences.

What are the important environmental factors that account for the wide individual differences in intelligence of which we are so painfully aware? What are the critical developmental periods in the person's lifetime when the ability to perform well on intelligence tests is established?

It is clear that the environmental factors that relate to the socioeconomic status of the parents of the child are of great importance. The damage done to the IQ of the child is not so much a result of inferior food, clothing, and shelter as of inferior aspirations and expectations. The poor do not have great expectations, as a rule, and for obvious reasons; they see no way to change their status significantly. Research results show that measured intelligence is determined early in life, before schooling has much chance to influence its development. Parents and peers during the preschool years are clearly major factors in intellectual development. We can ask how plastic intelligence is. Many cases of substantial changes in measurable intelligence have been recorded for people at various ages. Clearly it is possible to alter one's intelligence substantially by many different treatments. It is not useful to try to determine precisely the relative contributions to intelligence that result from environmental factors and from genetic factors. Nonetheless, the human brain is the source of our intelligence, and the brain depends upon its genes for its development in the same way that other physical organs do. The evolution of the brain is measured in millions of years, and a large number of gene loci had to be involved. Therefore a large number of gene loci were involved in the development of intelligence. It is clear from the study of Dewey et al [1965] that at least 100 gene loci are each capable of producing severe mental retardation. There are probably hundreds of other gene loci that also affect intelligence to a greater or lesser degree; clearly a multigenic situation.

We expect to be able to calculate heritabilities for multigenic traits, and that has been done many times for intelligence, with a wide range of answers. Some of the investigators who have calculated heritabilities used little common sense in their interpretations. In view of the plasticity of normal human intelligence, it does not make sense to claim that the variation in the trait has a heritability of 80 percent.

This value is too high, when considered in the light of the other knowledge we have, and should be considered less realistic than lower values. Intelligence must have a genetic basis, as do all important traits. What interests us most is the range of variation to be expected in the population and how much this can be manipulated by practical environmental changes.

Feldman and Lewontin [1975] think that heritability statistics related to intelligence are useless. I would agree with them that the heritability values for intelligence have not been of use in any practical sense and that they have been grossly misused in the discussion of racial differences in intelligence. Black children perform poorly on intelligence tests compared with white children. However, Nichols and Anderson [1973] and Rao et al [1977] showed that if the socioeconomic status for the two sets of children is similar, the performance on the tests is also similar. The first order of business is clearly that of improving the environments of the black children in the United States. We have the "vicious circle" problem in that in order to improve the children we have to improve the parents when they are children, and to improve the parents when they are children the grandparents would have had to have a better environment. These vicious circle problems take more than one generation to resolve, but that is no excuse for not working harder now. Progress has been made in the century since the formal abolition of slavery, and it is to everyone's advantage to speed up the process. Everyone should be able to develop to the maximum of his or her genetic potentiality.

The fact that the use of the heritability statistics for intelligence may have done more harm than good is no reason to reject all statistical concepts in relation to intelligence. Eaves [1973] provided an estimate of the narrow sense heritability of 60 percent for intelligence from the Reed and Reed [1965] data, which must be considered with caution as it can be applied only to their Minnesota sample. It tells us what we already know: that the genetic basis of measured intelligence is of importance and that the environmental basis is likewise of importance, at least in Minnesota.

Rao et al [1976] have used the method of path coefficients to partition the variance in IQ expression and provide the same kind of distinction found for diabetes; namely, the genotype of children accounts for more of the phenotypic variance (0.670 ± 0.067) in IQ than does the genotype of parents (0.211 ± 0.104). They make the interesting statement that, "since the family environment is so important, it is conceivable that adult education of parents could, by diminishing the intergenerational path between family environments, have greater effects on academic performance than preschool education of their children." They also state that, "whereas family resemblance of children is largely genetic, for adults it is largely due to their childhood environments, presumably acting on occupational aspirations."

We must consider heritability values to be open to various uncertainties, but we should not abandon the value of the concept, because there must be a genetic basis for intelligence. No matter how plastic and malleable intelligence may be, we know that all chromosomes (except one X) must be present for a person to have normal intelligence. We know many specific gene loci at which at least one normal allele must be present in order to have normal intelligence. Twin studies indicate the presence of a genetic basis for intelligence, as do numerous adoption studies, many of which are very recent and sophisticated in design. Among these are the study of Munsinger [1975] on children's resemblance to their biological and adoptive parents in two ethnic groups, that of Horn et al [1979] on the Texas adoption project, and finally the study of Scarr-Salapatek and Weinberg [1975] on black and interracial children adopted into above-average white homes. All of these studies show that both the child's genetics and environment contibute significantly to his intelligence and, presumably, in somewhat different proportions for each child.

We will consider the genetics of mental retardation in the next chapter, as it is one of the more perplexing problems for the physician. Our final comments about intelligence in this chapter concern the gifted child; such children have problems, but of less gravity than those of retarded children. The gifted have problems because of their high intelligence, which can separate them from the average population.

After 1850, or thereabouts, the precocious child tended to be classified as abnormal, as neurotic and headed for insanity. The idea grew that the bright child should be protected from intellectual stimulation, and that any tendency toward early cleverness should be discouraged. "Early ripe, early rot" was the slogan of those who favored slow maturation. This attitude seemed to be rather superstitious to Lewis Terman, who carried out a vast screening program in California to select the 1,000 children with the highest IQs and follow them through their lifetimes. This was done, and the spouses of the gifted individuals who married, likewise were studied carefully. The spouses were not as highly intelligent as the gifted individuals, but they had an average IQ of 130 compared with the average IQ of 150 for the gifted persons. The average of the gifted and their above-normal spouses was 140, of course. The children of these unusually fortunate parents were tested also and had an average IQ of 130, or ten IQ points less than the parental average of 140. This drop of the children's IQ of ten points is called Galton's filial regression toward the mean — that is, toward the normal population average of 100 IQ points.

The drop of ten IQ points of the average for the children of the gifted persons does not seem remarkable. The problem is that the individual variation in IQ was great though no greater than one might expect for a multigenic trait. There were 1,538 children tested, and 13 were below IQ 69 and therefore mentally retarded.

There were 37 children with IQ values between 70 and 99, and another 101 children with IQ values between 100 and 109. These 151 children with low or mediocre test scores would often be a source of frustration and disappointment to their gifted parents. The parents will react much better to the situation if they are informed that many of the children of extremely bright parents will regress toward the average IQ. If on the one hand, the gifted parents expect their children to be equally gifted, and they are not, much psychological damage to both the children and the parents may result. On the other hand, if the parents are prepared for some of their children to have lower intelligence, they can make a good adjustment to the individual differences of their offspring. The goal is to help all children develop to their fullest potential. The genetic potential for each child is different from that of every other one, and this axiom should never be forgotten. Furthermore, the child who is not as bright as his parents had expected him to be may well exceed his parents in other talents, such as generosity and affection. These latter traits should be cherished.

18
Mental Retardation

Many parents of mentally retarded children have gone from one physician to another and have been told that their child "would grow out of it" and other such evasive comments. "Doc" thinks he is doing the parents a kindness by letting them continue to hope that their child is not retarded. Fortunately, the properly trained physician knows that the parents must eventually accept the facts about their retarded child. The earlier a correct evaluation of the situation is provided, the better it will be for all concerned. The physician should be understanding and compassionate toward the parents and try to alleviate their anguish. The physician must be sure of his diagnosis and not plague the parents with unfounded speculation.

Most of the traits that intrigue the human geneticist are rare, such as many of the inborn errors of metabolism that the physician will seldom encounter in his practice. But the traits described in this chapter are far from rare. We are concerned here with those that were known in the past as feeble-minded, then as mentally deficient, then as mentally retarded, and, finally, as exceptional children. This evolution of terms is obviously an attempt to soften the stigmata that go with any label for a disadvantageous trait. Whatever label we use, mental retardation is an extremely frustrating condition, because those affected are not ill in the usual sense, nor do they die from their anomaly, as a rule. Their main problem is that their intellectual development is slower than that of their siblings and peers. Physical development is sometimes retarded, but the cardinal feature is the retardation of the mental progress expected in our society.

Parents find it difficult to accept the fact that mental retardation may have a large genetic component. They think of heredity and mental retardation in terms of gruesome families of the Jukes and Kallikak type. However, such socially degenerate families are at one extreme of the many types of mental retardation, and there is no reason whatever why most parents of retarded children should erroneously identify with such social problem kinships.

There are more than 100 gene loci that can result in some form of severe mental retardation, as demonstrated by Dewey et al [1965]. Chromosome anomalies are an important cause of mental retardation (see Speed et al [1976]), and environmental insults can also be primary causes of mental retardation. It seems almost ridiculous to include over 100 distinct and different genetic types of mental retardation as one trait. The only reason for doing so is that many of the types are not visibly different, even though they are genetically distinct. Further-

TABLE 18-1. The High Concordance of Identical Twins With
Mental Retardation Compared With the Lower Agreement
Between Fraternal Twins

	Both mentally retarded	One only mentally retarded
Identical	74 (97%)	2
Fraternal	80 (37%)	138

more, the whole collection can tell us some things about mental retardation even
when so many genetically different types are combined into one "trait." This is
illustrated by the comparison of concordance for mental retardation, as a "combined trait" in identical and fraternal twins shown in Table 18-1.

Larger sample sizes and modern diagnoses might alter these percentages somewhat, but the general observation of substantially greater concordance for one-egg
twins compared with two-egg twins would remain. The fraternal twins are, of
course, full siblings, and much of their concordance will have a genetic basis of
the simple mendelian recessive type. We shouldn't forget the effects of the environment in contributing to mental retardation, particularly of the "familial"
type. In the "familial" type one observes that one or both parents and several
of their large family of children fall slightly below an arbitrary value of 70 on an
IQ test. A substantially better environment might well have prevented any of the
members of the family from being scored as retarded.

There are numerous extensive pedigrees in the literature showing the inheritance of X-linked nonspecific mental retardation, such as the one reported
by Yarbough and Howard-Peebles [1976]. In this case there was a seven-genera-
tion pedigree with 19 known affected males who appeared to have received the
gene through their normal mothers. Retardation, lack of fine motor coordination,
hyperactivity, and a speech defect were the characteristics of these affected
males. In such cases the environmental contribution seems to be less crucial to
the outcome than the X-linked gene. (See also the large study by Pauls [1979].)

The usual experience for the physician will not be this clear. Ordinarily the
only information available will be from the retarded child and his immediate
family. If the child cannot be placed in a specific etiological category he can be
assumed by exclusion to fall into the multifactorial group. Empiric risks are our
only way of predicting expectations of mental retardation for such children.
Table 18-2 was arranged by Nora and Fraser [1974] from the data of Reed and
Reed [1965]. It can be seen that for any normal couple without retarded close
relatives, the chance that any child of theirs will be retarded is less than one percent of the cases. However, if both parents are retarded and they already have a

TABLE 18-2. Risk of Mental Retardation in Children and Sibs of
Retardates (IQ less than 70)

Parents	Number of retarded Children	Risk for	Risk (%)
0	– – –	child	1
1	– – –	child	11
2	– – –	child	40
0	1	sib	6
1	1	sib	20
2	1	sib	42

retarded child, the risk that the next child will be retarded is about 42 percent. The latter risk is a serious one, and in many cases the county welfare board will intervene to prevent the birth of any more children. The physician should become involved in such problem family situations, as he has a responsibility to the community as well as to his patients. He is the one who not only has to attempt a diagnosis and prognosis of the retardation but also should provide the risk figures for retardation in possible subsequent children.

One of the difficult questions asked by people who wish to adopt a baby will be whether it is likely to be mentally retarded. Ordinarily, agencies would not place a baby for adoption if it had Down syndrome or some other obvious type of mental retardation. However, three-quarters of retarded infants are undifferentiated and are not identified until a few months or years after birth.

There are numerous studies comparing the intelligence of adopted parents. The children were tested after having lived with the adoptive parents for many years. The study of Skodak and Skeels [1949] on the IQs of adopted children and their biological mothers and adoptive parents is unique in that individual IQ scores are available for each child and its biological mother. Thus, there were 11 biological mothers who were mentally retarded, with an average IQ of 63. Their 11 children who were placed for adoption scored an average IQ of 94, a gain of 31 points. This is not as remarkable as it might seem, as their biological fathers IQ values would have had to be higher than 63. Because of Galton's law of regression toward the mean, the children would have averaged higher than the 63 of their mothers even if they had remained with them instead of being placed.

The important point is that the adoptive parents were not adopting an average, but rather an individual child. The bottom three of the 11 children from retarded mothers had IQs of 66, 74, and 87. Thus only one of the 11 was retarded by definition, but all three were too low to show satisfactory intellectual growth in the type of family that is given foster children. The difficult question for the agency and their consulting physician is whether all children from retarded mothers should be denied placement until old enough to give reliable intelligence test results in order to protect the one-fourth of the adoptive couples from getting children whose IQs would be unsatisfactory.

Presumably the agency will consider each baby of a retarded mother on an individual basis. One of the most important points to consider will be the intelligence of the biological father of the child.

The physician has many important social service functions connected with mental retardation. The parents of a retarded child will usually ask their doctor's opinion about whether they should place the child in a boarding home or state institution. Such decisions are very hard to make, and the physician's input is of crucial importance. In the institution the child will find himself among his equals and he will be able to react with them successfully, whereas in the home community he is likely to be overprotected from social contacts or cruelly rejected in his social efforts. The retarded child can easily disrupt the family behavior pattern and can be exploited by others. Nonetheless, sometimes it is worth taking these risks and maintaining the retarded child at home. The main difficulty is that of helping the parents to find a realistic stance in regard to the future of their retarded child or children.

There has been a vigorous move in recent years to "normalize" the retarded and return them to the community either singly or in group homes. There has been the mistaken notion that it is financially cheaper to maintain the retarded in the community in order that a large state institution might be closed. This is economic folly, because supervision of an individual or small group has to be more expensive, if adequate services are provided, than when administered in larger units. Living in a normal community is not necessarily beneficial to all retarded persons.

The physician must have some concepts for action in these social areas, because about 3 percent of all the newborns will be mentally retarded. Fortunately, he can obtain ideas concerning his local facilities from the county or city association for retarded citizens.

The enormity of the mental retardation problem led to great public concern and accounted for the rise of the eugenics movement and the passage of sterilization laws in many states. Such laws were well-intentioned, and it was hoped that the frequency of mental retardation would be lowered. If there were fewer retarded persons, better care and attention could be given to each of them. The saying that hell is paved with good intentions is appropriate here, as some of the eugenics activities were not beneficial to some of the recipients. The public concern arose because the dire poverty and degenerate social conditions of retarded persons who produced large numbers of retarded children were obvious to everyone. It was reasonable then to assume that the retarded were out-breeding the rest of the community and that the national intelligence must therefore decline.

Higgins, et al [1962] showed that in actuality, while there are some large "social problem" families with much mental retardation, many retarded persons have no children at all (either because they are under supervision or because they are too poorly adjusted to obtain a mate). Thus, when all of the retarded persons in any generation are studied, it is found that they have fewer descendants than do the normal members of that generation. Because of early death, institutionalization, incapacitation, and the other disadvantages of their defects, the mentally retarded are at a disadvantage in relation to normal people in regard to reproduction when everyone in a particular generation is included. However, new mutations and chromosome aberrations continue to supply new cases of retardation, so that mental retardation will continue to be one of our major social problems (see also Reed and Reed [1965]).

The normal parents of retarded children are greatly concerned about the possible reproduction of their children, normal as well as retarded. The primary question is, should the retarded marry? Marriage is a valuable foundation for a person's self-esteem. Children add to the feeling of worthiness of the parents. However, if the marriage is a failure, and the children are abandoned or neglected, the situation is not favorable to self-esteem and a feeling of worthiness. Thus the physician may become involved by providing temporary or permanent contraception for the young retarded adult or adolescent.

Reed and Anderson [1973] used the Reed and Reed [1965] data to show that both the proportion reproducing and the risk of retardation in the offspring were higher for retarded females than for males. The net number of retarded children who survived infancy was three times larger for retarded mothers (0.31) than for retarded fathers (0.10). There is no need to speculate about the causes for this well-established difference; the point of interest is that decision-making is more crucial for retarded girls than for retarded boys as far as reproduction is concerned. Administrators are inclined to forget that in the "normalization" process for the retarded the likelihood of increased heterosexual activity is real and significant. Clearly, contraceptive supervision is more important for the retarded than for normal young people, as the consequences of inappropriate reproduction are more serious; the risk of producing retarded children is many times greater than for normal persons.

There are many other problems for the physician in the area of mental retardation, particularly those involved with ethical questions. For instance, should open-heart surgery be offered to children with Down syndrome and heart defects? How heroic should the attempts be to extend life a few more days for a child with Tay-Sachs disease, as early death is a certainty at present? These are important ethical decisions, but they are not part of genetic counseling as there are no genetic consequences, whatever the decisions. On the other hand, decisions are made every day for less severely affected retardates, and these usually do have genetic consequences. They have extremely important effects on the family and community concerned, even though they may not affect the worldwide gene pool appreciably. The mental deficiencies and the mental illnesses are our most important and sensitive health problems, and much more attention and money must be devoted to all aspects of their solutions.

19
The Psychoses

Mental disorders represent the last line of resistance to genetics. Psychiatrists have little choice but to accept the clear-cut dominant mendelian pedigrees for neurological disorders such as Marie cerebellar ataxia and Huntington disease, but some are emotionally unprepared to accept the idea that heredity has anything to do with the psychoses. This is strange, because Freud was well aware that there is a constitutional (genetic) basis for mental illnesses. Some psychiatrists reject the concept of a genetic basis for the psychoses because they have the idea that genetic diseases cannot be cured. If the patients believe in heredity, which they usually do, such thoughts might retard their therapy; actually such ideas are not relevant to proper treatment of the psychoses.

The main reason for the disarray of thinking among most persons concerned with the psychoses is simply our lack of scientifically tested data and concepts relating to many aspects of the psychoses, including the diagnoses. It is hoped that a few of the ambiguities will be clarified in the following pages.

A correct diagnosis is desirable for the treatment and proper management of any patient. Fortunately, many patients get better even when they are given an incorrect diagnosis and the wrong treatment. The taxonomic problem is unusually acute for the mental disorders, because even when it is correct, the diagnosis may have to be changed to a different one, which may also be correct, at a later time in the life of the patient.

There are psychologists and psychiatrists who seem to be confident of their ability to make the correct diagnosis in each case, but they often disagree with the equally confident diagnoses of their peers. The reader is referred to an intriguing little paper by Nathan, et al [1969] as an illustration of the diversity of diagnostic labels given to one patient when observed by 32 health professionals at the

Harvard Medical School. There was no question but that the 36-year-old man was mentally ill. Two of the observers declined to make a diagnosis, but the remaining 30 professionals conferred 14 different diagnostic labels which ranged from paranoid schizophrenia to "depressive reaction and temporal lobe epilepsy." Some psychotics display only the classical symptoms of a schizophrenia or of manic-depressive psychosis, but most patients display behavior that fluctuates, with symptoms of both schizophrenia and manic-depressive psychosis, plus damage due to alcoholism or to other related mental difficulties. The final diagnosis is the result of the importance a particular psychiatrist attaches to the set of symptoms that seems most significant to him as a result of his training and his own mental set. The same patient can recieve a different diagnosis from a different psychiatrist because the second physician had a different training or outlook on life or because in the meantime the patient's symptoms have changed.

The gravity of the taxonomic problem has been recognized by the studies of the United States – United Kingdom Cross-National Project. This extensive endeavor included a section on videotape diagnostic interviews that were shown to fully qualified psychiatrists in Great Britain and the United States. Many more of these cases were classified as schizophrenia and fewer as affective disorders in the United States than in London. As the tapes were the same for both audiences, the resulting diagnostic differences were obviously due to the differences between the American and British pyschiatrists. We cannot spend more time on this important problem here, but an interesting summary of the Cross-National Project can be found in the article by the project staff (Zubin et al [1974]).

The problem of diagnosis is particularly difficult for genetic studies, because some important studies such as Kallmann's [1938] find only schizophrenic relatives of a schizophrenic proband or only manic-depressive relatives of a manic-depressive proband. His "misfits" in the schizophrenia families were labeled as "schizoidia." More recent studies such as that of Ödegård [1972] , show a correlation between the diagnosis of the proband and that of his relative, but it is far from unity. Table 19-1 is my condensation of Ödegård [1972] table showing the modest, but far from perfect, correlation between the diagnosis in the proband and that of the psychotic relative. The substantial discordance between diagnoses of proband and relative is fatal to a single-gene hypothesis but is in good accord with expectations on a multigenic hypothesis. Our own very large study (Reed et al [1973]) showed large numbers of differences of diagnosis between relatives. In our study, like the others, relatives often had "other mental disorders" which are related to the illness of the proband in some way. These "other mental disorders" can be explained on a multigenic basis but not easily on any single-gene locus theory. They must be explained in some way, as they are more frequent than the more clear-cut psychoses of the schizophrenic or depressive types.

TABLE 19-1. Disagreement Between the Diagnosis of the Proband and His Relative (condensation of Ödegård's Table 11-IV)

Diagnosis in the probands	Schizophrenic relative	Reactive psychosis relative	Affective relative	Total
Schizophrenic proband	426 (375.8)	107 (129.0)	123 (151.3)	656
Reactive psychosis proband	23 (47.0)	39 (16.1)	20 (18.9)	82
Affective proband	23 (49.3)	16 (16.9)	47 (19.8)	86
Total	472	162	190	824

$$\chi^2 = 112.1; d/f = 4; P < 0.001$$

*The numbers in parentheses are the expectations if there is no association between the diagnosis in the proband and his psychotic relative.

From Kaplan: Genetic Factors in Schizophrenia, 1972, p 267.

McCabe et al [1972] point out that there is a "very great overlap between good prognosis schizophrenia and schizo-affective disorder." The more seriously ill relatives are to be found toward the schizophrenic pole and the less seriously ill at the affective disorder pole.

It is general knowledge that when diagnoses are changed for patients, the change is usually from manic-depressive psychosis to schizophrenia and seldom in the opposite direction. When the diagnosis changes, it is in the direction of a more serious illness, schizophrenia. Clark and Mallett [1963] carried out a follow-up study of schizophrenia and depression in young adults. Though they were followed for only three years, the schizophrenics retained that diagnosis in 93 percent of the readmissions, whereas one-third of the depressives were rediagnosed as schizophrenics. Conversely, Abrams et al [1974] rediagnosed 41 paranoid schizophrenics classified by other psychiatrists leaving only two cases as schizophrenics.

We have an unusually rich literature concerning twins and psychoses. Concordance is surprisingly high (to me) in view of the nature of the trait and the usual difficulties of diagnosis. Gottesman and Shields [1966] realized the necessity of including other mental disorders in viewing the co-twins of the probands. One can see in Table 19-2 that only 42 percent of the genetically identical twins received a diagnosis of schizophrenia for both members of the pair. However, another 37 percent of the cotwins were treated for some kind of mental problem. The final 21 percent were not treated for any mental problem and were considered normal, although they, of course, had the same genetic potential for a mental problem as the the 79 percent of the co-twins who were treated, as well as all the probands who were schizophrenic patients.

TABLE 19-2. Change in Concordance With Increase in Breadth of Diagnosis (Adapted from Gottesman and Shields, 1966)

	Monozygotic pairs		Dizygotic pairs	
	Number	Percent	Number	Percent
Both twins hospitalized for schizophrenia	10	42 ⎫	3	9 ⎫
		⎬ 54 ⎫		⎬ 18 ⎫
Second twin with other psychiatric hospitalization	3	12 ⎭ ⎬ 79	3	9 ⎭ ⎬ 45
Second twin with treated personality disorder	6	25 ⎭	9	27 ⎭
Second twin normal	5	21	18	55
	24	100	33	100

There is a most fascinating set of identical quadruplet women concordant for schizophrenia who were studied from birth – not with the anticipation that they would develop schizophrenia, but because quadruplet births always attract attention. As the years went by, they and their weird parents were studied by Rosenthal [1963] and collaborators at the National Institute of Mental Health in Bethesda, Maryland. It was impossible to determine any age of onset for their schizoprenia, even though the evidence was detailed and voluminous. The girls were different from other children at an early age and gradually became more so. In Rosenthal's words, "The girls seemed to have had characteristics from a very early age that may best be described as constrictive. They had low energy levels, were rather placid, 'sweet,' introversive, not very talkative, communicative, or outgoing, and they took little initiative either with respect to general activities or interpersonal relationships. On top of this behavioral pattern was imposed a severely constrictive family regimen, their parents transcending all bounds of reasonableness in this regard. Environmental constriction was heaped upon inherited constriction. When the girls became ill, the predominant feature of their illness was extreme constriction as manifested by catatonic symptoms."

Another interesting point about these Genain quadruplets is that the severity of their illness was directly correlated with their preference by their parents. The smallest one at birth was the least favored and the most seriously ill, with permanent hospitalization. The most favored girl recovered from her illness sufficently to marry and have a child. All four had the same genetics and were obviously schizophrenic, but the range and severity of their symptoms varied greatly.

The average concordance of about 50 percent for the psychoses in identical twins is higher than that for most of the common congenital malformations such as as clefts of lip and palate (38 percent) or pyloric stenosis (21 percent). This may not be because of a stronger genetic component for the psychoses than for pyloric stenosis, for instance, but instead may be due to the much longer environmental association between the twins having a psychosis than those with a congenital malformation, which is present at birth. Incidentally, the heritability for pyloric stenosis was estimated to be 79 ± 5% by Falconer [1965].

There is also an excellent twin study of manic-depressive disorders by Bertelsen et al [1977]. Among the co-twins of 69 monozygotic probands there were 46 with manic-depressive disorders and a further 14 who had other psychoses or marked affective personality disorders or had committed suicide. These are concordance rates of 67 and 87 percent, respectively; the dizygotic twins had comparable concordance rates of 20 and 37 percent, respectively. These rates are evidence of a strong genetic basis for the affective psychoses.

Very convincing evidence for a genetic basis for the psychoses is the adoption study of Heston [1966]. His experimental subjects were born between 1915 and 1945 to schizophrenic mothers confined to an Oregon psychiatric hospital. All the babies were discharged from the state hospital within three days of birth to the care of family members or to foundling homes. Children were not included if they were released to maternal relatives, but children released to paternal relatives were accepted. None of the children were released to suspected schizophrenogenic environments. The starting number of 74 babies ended with only 47 experimental subjects, mainly becuase 15 children died before reaching school age. Fifty control subjects were selected via the records of the same foundling homes that received some of the experimental subjects.

The schizophrenic mothers produced five children who became schizophrenic, while none of the control children became psychotic. This is 10.6 percent schizophrenic children, without age correction, which is as high a frequency of psychotic offspring as found by Kallmann or any other researcher where the children were reared by the psychotic mother. Karlsson [1966] and Reed et al [1973] had studied families in which the children of a schizophrenic parent were adopted and other families where the children were raised by the psychotic parent. Later there was a higher frequency of schizophrenia among the children who were placed for adoption than among those who were raised by their schizophrenic mothers. Our explanation of this unexpected finding was that the children of the most severely ill mothers had to to be placed for adoption and that the most severely ill mothers had a heavier genetic load for psychosis than did the schizophrenic mothers who were able to raise their children.

The various adoption studies all seem to agree in eliminating environmental factors from having fundamental significance in the etiology of the psychoses. The genetic potentiality for a psychosis is present at conception. The environmental factors contribute to the severity of expression of the disorder and are essential for the onset of each psychotic episode. Probably the failure to identify specific environmental factors, except such generalities as "social class," which is clearly important (see Reed et al [1973]), results from the fact that the genetic load for a psychosis is slightly different in every individual, and each potential psychotic reacts to each environmental fluctuation in a different and unique way (see Weissman [1979]).

A most ingenious series of experiments has been carried out by Rosenthal et al [1975] with four groups of children, including adopted and natural children, with or without a psychotic parent. The degree of illness in the child was correlated with the quality of parent-child relationship. Their conclusions were that "rearing patterns have only a modest effect on individuals who harbor a genetic background for schizophrenia, but an appreciable effect on persons without such a background." Unfortunately, this statement does not provide evidence of how environmental factors contribute to the substantial discordance in the behavior of identical twins.

Let us turn to the practical questions of the empiric risks of a psychosis for siblings and children (first-degree relatives), nieces and nephews (second-degree relatives), and first cousins (third-degree relatives) of a psychotic proband. The data in Table 19-3 are selected from various sources.

There are five different types of siblings shown in Table 19-3, starting with the genetically identical twin sibs and ending with the unrelated step-sibs (often adopted children). All five types of sibs were raised in the same households with

TABLE 19-3. Frequency (Percentage) of Psychoses in the First-, Second-, and Third-Degree Relatives of Schizophrenic or Manic Depressive Probands

	Genetic correlation	Schizophrenic proband	Manic depressive proband
Identical twin siblings	100.0	55.0	65.0 (?)
First-degree relatives			
Fraternal twin siblings	50.0	14.0	25.0
Ordinary siblings	50.0	17.8	21.5
Children	50.0	16.0	12.0
Second-degree relatives			
Second-degree relatives			
Half-sibs	25.0	7.0	16.7
Nieces and nephews	25.0	10.0	—
Third-degree relatives			
First cousins	12.5	3.0	—
Nonrelatives			
Step-sibs	0.0	1.8	0.8
Control groups	0.0	2.0	2.0

the proband in most cases. There is no claim that their environments were the same, but they could be expected to vary in a somewhat random fashion, with the step-sibs perhaps being given the "worst" treatment. If so, the "worst" treatment was good for them, because they had no more mental disorders than the control populations. There is, on the other hand, an extremely high association between the frequency of psychoses and the genetic relationship of the sibling to the proband. This correspondence between genetic correlation and frequency of a psychosis in the relative of the proband is so orderly and so highly significant that the importance of a genetic basis for the psychoses cannot be denied. At the same time, the fact that even the identical twin sibling is often discordant from its twin proband shows the striking importance of some kinds of environmental events necessary for the expression of the genetic potential for a psychosis.

Except for the identical twin siblings, none of the groups of relatives of the proband have half as many psychotic members as expected from the genetic correlation — that is, the percentage of genes that the relative and the proband have in common. My estimate from the rough data in Table 19-3 is that about one-third of the people who have the genetic potential for a psychosis do become sufficiently ill to receive medical treatment and to come to the attention of the researcher doing a family study. At least an equal percentage of the relatives also appear in the statistics as having "schizoidia" (Kallmann [1938]), "minor mental deviations" (Odegård [1972]), "other mental disorders" (Reed et al), or some other designation of the authors' choice. The "other mental disorders" include severe neuroses, personality disorders, and chronic alcoholism in all studies where the relatives are studied closely.

The person who is awarded the label "psychotic" is *not* given the label as a result of being three standard deviations away from some average performance of a "normal" population, but because of some decisive event such as admission to a mental hospital or treatment by a psychiatrist. Another person with similar behavior might escape the psychotic label by evading hospitalization or treatment. However, this person who has evaded the psychosis label would receive that of "other mental disorder" in a proper study. The real diagnostic difficulty comes with the dichotomy between "other mental disorder" and "normal." Probably all family studies that are carefully done end up with the realization that these dichotomies are arbitrary, and though useful, are not truly descriptive of the population. All the evidence indicates that there is a continuous range of mental illness and that it is a substantial component of a normal curve of human behavior.

Hegnall [1966] made a careful study of an entire Swedish population and showed that the estimated cumulated risk of developing mental illness up to 60 years of age was 43.4 percent for men and 73.0 percent for women (those who had at least consulted a physician about their mental health). His estimated cumulated risk of severe mental impairment up to 60 years of age was 7.9 percent for men and 15.4 percent for women. His definition of mental illness is perhaps rather broad, but it is tenable and demonstrates very clearly the normal curve characteristics of human behavior.

We now come to the question of the genetic mechanism or mechanisms involved with a normal curve type of trait with continuous variation. It is quite possible that there are many discrete types of mental disorder with single gene locus inheritance included in the large percentage of persons in a population with a mental illness. This is the case for the normal curve of intelligence; here there are simple recessive traits like phenylketonuria and galactosemia, where the retardate can be identified by chemical screening tests. Huntington disease includes some mental illness in most cases and depends upon a single gene locus for its behavior as a mendelian dominant gene. However, no successes have yet resulted from attempts at biochemical screening for the usual psychoses. If the psychoses are merely a component of the normal curve of behavior, then one would not expect to find qualitative chemical differences, only quantitative continuous gradations with no natural cutting point. Furthermore, no single gene locus would be involved, but instead a large number of gene loci would contribute to the continuous variability. The latter is what one might expect for a trait with obvious adaptive importance that has probably been of significance for millions of years and perhaps can be studied in our distant relatives, the chimpanzees, because of their great biochemical and chromosomal similarities to us.

It was thought by some that lithium had an action that was specific to the manic behavior of manic-depressive patients. However, it turned out that the antiaggressive effect was a general, rather than a specific, effect and resulted in improvement in some cases of aggressives with phenylketonuria (see Worrall et al [1975]).

A great deal of interest has developed recently in the possibility that manic depressive psychosis depends upon an X-linked dominant gene (see Winokur and Tanna [1969] and Mendelwicz and Rainer [1974]). It is not easy to distinguish between sex influence, where there is an excess of affected females, and an X-linked dominant gene. All the studies included families that showed father-to-son transmission, which eliminated X-linkage for them. Coincidence could account for some of their pedigrees where X-linkage did seem to be present. At any rate, X-linkage is still not established as an important factor for depression as a whole, in my opinion, as well as that of Loranger [1975] and Gershon and Bunney [1977]. In the reports by the latter authors it is shown in several ways that the association between bipolar affective disorder and color blindness is not likely to be the result of chromosomal linkage.

Attempts to determine whether one gene locus or many gene loci are involved with the etiology of a particular psychosis have not been successful so far. It seems more reasonable to me to assume that more than one gene locus must be involved with a frequent psychosis such as schizophrenia. If schizophrenia does have a multigenic basis, then the methods used in quantitative genetics should be helpful.

In order to measure the variation of the trait quantitatively — ie, how much of it, relatively speaking, is due to genes and how much to environmental factors — one can calculate the heritability. This was done for schizophrenia by Gottesman and Shields [1967]. They found that the heritability for various degrees of relatives of a proband was of the order of 65–85 percent, regardless of the degree of relationship to the proband. Thus, the roles of genetics and environment were about the same in the aunts and uncles of the proband as in his siblings. These heritability values can be only rough approximations to reality because of the nature of behavioral traits. However, the heritability of 65–85 percent indicates that more than half of the variation in relatives of the proband as to whether they were psychotic or or not was due to gene differences and less than half was due to differences in environmental factors.

Thus a large amount of research has turned up no evidence of any important type of psychosis being due to a single gene locus. Perhaps there are large numbers of individually rare types of psychoses, each of which depends upon simple mendelian transmission at one gene locus; however, until such conditions have been identified, we will have to use the empiric risk figures shown in Table 19-3 for genetic counseling purposes. That is, if a parent is psychotic there is a chance of roughly 15 percent that each child would develop a psychosis before old age. If both parents have a psychosis we can expect some 40 percent of the children to develop a psychosis before death. These two values will serve for both schizophrenia and manic-depressive psychosis, as the statistics for the two disorders are fairly similar.

There is no way of determining the true risk for a psychosis in a particular individual, so we have to use the risks determined from family studies. The clients are not interested in the precise percentage risk given to them, but rather whether it is high, moderate, or low. Furthermore, while genetic counseling for mental disorders is of considerable social interest, we have never had a large demand for it at the Dight Institute. There has been a large demand for counseling for Huntington disease, which is a rare trait, but less demand for counseling for the psychoses, which are frequent. Perhaps this difference is due in part to the fact that the genetics of Huntington disease is well known, whereas the genetics of the functional psychoses is not; furthermore, it is the congenital disorders that come for counseling most often. However, I don't wish to give the impression that genetic counseling for the psychoses is not important. It is, and an especially interesting paper by Erlenmeyer-Kimling [1976] is well worth reading.

The long paper by Kay [1978] on an assessment of familial risks in the functional psychoses and their application in genetic counseling is not only timely but also covers much of the same material summarized in this chapter.

Finally, I wish to recommend the very short essay by Shields [1977] titled "The Major Psychoses." It discusses most of the points considered in this chapter and does so with a beauty and clarity that make it well worth reading.

20
The Environment

No genotype can be expressed in a vacuum. The study of genetics must always attempt to evaluate environmental variables because they can confound any genetic interpretation so easily. One can never really distinguish between heredity and environment, because the interaction between them is continuous from the womb to the tomb. However, one can single out a specific gene causal to a defect and demonstrate its Mendelian mechanism of transmission within each family pedigree. We are not only the products of our heredity and environment, but as H. J. Muller has stated, "we are the genes' way of making more genes."

The geneticist is intensely interested in the environmental factors that confuse the genetic picture and is delighted when an environmental factor can be pinned down. The relative importance of environmental factors will vary from one trait to another. Some traits will have an important environmental origin, others will be little affected by environmental fluctuations, but the great majority of human differences will result from complicated interactions between multiple genetic and environmental factors.

The sickle cell gene relationship to malaria research is one of the most brilliant triumphs in genetics. The original mystery was the question of why the gene for sickle cell anemia was practically nonexistent in northern Europe while it was present in about 40 percent of the members of some African tribes. Allison [1954] showed that the sickle cell gene in one dose provides significant protection against malaria so that where the malarial parasites abound the gene for sickle cell anemia will become more frequent, even though it may be lethal on its own account in the double dose or homozygous condition. The parasite is the environmental agent, and the local populations gradually evolve a genetic protection against it. In the United States, where malaria has been eliminated, the sickle cell gene is probably being reduced in frequency as it no longer has a protective function.

The sickle cell gene is our best documented example of heterozygous advantage of a human gene. Li [1975] has calculated that a reasonable range for the heterozygous advantage seems to be from 5 to 18 percent. Other human recessive genes which have higher frequencies in the heterozygous condition than would be expected from their mutation rates, such as those for cystic fibrosis and Tay-Sachs disease, have not been linked to any adverse environmental factors, and their frequencies may be merely the result of genetic drift.

Numerous retrospective studies in which affected children have been reportedly damaged as a result of anoxia have generated widespread fears and fancies that that intelligence later on will be lowered severely. Such reports are usually biased by the selection of cases so that the unaffected babies are not included in the sample or there are other statistical defects in the study. However, the careful and long-time follow-up study of Keith and Gage [1960] with a Mayo Clinic sample showed a mean IQ of 116.8 for the children from the prolonged labor cases, 117.3 for the asphyxia cases, and 114.3 for the controls. The differences are not statistically significant or suggestive of any effect from either prolonged labor or anoxia. No one would claim that anoxia is good for a newborn baby, but it seems to be somewhat uncertain as to what long-term damage is caused by it.

The topics to be taken up in the sub-sections of this chapter share the quality of being logical environmental agents for causing gene mutations or chromosomal aberrations. The matter of direct proof that any one of them has caused genetic damage in people is another matter. We will see what evidence can be brought to bear on this very important and practical subject. We will not consider obvious environmental defects such as lead poisoning or retrolental fibroplasia as it is not expected that the excess lead or oxygen are behaving as mutagens.

RADIATION AND NUCLEAR POWER

H. J. Muller won his Nobel Prize [1946] for work done in the 1920s, which demonstrated that an environmental agent (x-rays) could cause gene mutations in the chromosomes of *Drosophila.* His repeated attempts to convince physicians that their use of medical radiology often was unnecessarily hazardous, were not well received. One of the beneficial results of the atom bombing of Hiroshima and Nagasaki was that it shocked physicians both into greater safety efforts in their own radiology facilities and into an interest in human mutations induced by x-rays —and from there into an interest in the whole discipline of human genetics. The bomb blasts in Japan heralded the advent of modern medical genetics.

The United States government provided substantial financial support for studies on those receiving irradiation from the bombing, which produced vast amounts of data of various kinds but little or none demonstrating any *genetic* effects of the radiations. One would assume that many chromosomal aberrations and gene mutations were produced by the massive radiations. Why were these not detected? Presumably the gross chromosome aberrations had been eliminated by selective cell survival. The majority of gene mutations would be expected to be recessive in nature and therefore concealed in the Japanese genotype for generations to come. One of the important but unexciting conclusions from the vast effort was that, regardless of whatever genetic effects may have resulted from the radiation, their products or anomalies were indistinguishable from those already being produced in the Japanese people.

Neel et al [1974] reported on their continuing study of mortality rates among children born to survivors of the atomic bombings and a suitable group of controls. The average interval between birth and verification of death or survival is 17 years. The children were conceived within one month to 13 years after parental exposure. One of the possible manifestations of exposure to ionizing radiation is a decrease in life expectancy of children born to radiated parents, because of the induction of deleterious mutations in the gonadal tissues of these parents. In this study one or both parents were supposed to be within 2,000 meters of the hypocenter at the time of bombing. In spite of meticulous data collection and analysis there was no statistical evidence of any result of the bombing on the mortality rates of the children. The radiated parents received jointly an estimated dose of 117 *rem*. It *was* possible to demonstrate damage to fetuses being carried at the time of the irradiation, as one might anticipate.

The failure to demonstrate effects of radiation in children conceived *after* their parents were radiated by the atomic bombing in Japan does not prove that there were no effects. Certainly recessive mutations should have been produced in the irradiated persons, but there is no practical way of detecting them in their offspring subsequently conceived.

The obvious question arises as to why we should expect genetic damage to be detected in the offspring of medical personnel when there was no evidence in the children of people who survived an atomic bomb. The answer is not clear because we do have direct evidence that there are more affected offspring of medical personnel or of irradiated patients. We have previously cited the paper by Cox [1964] showing that there were more abnormalities among the offspring of mothers who had received extensive diagnostic irradiation for their cases of congenital hip disease. There were 7.5% severely affected offspring of the patients and 2.5% severely affected offspring of the controls. The difference is statistically significant and cannot be ignored. The irradiation is directed toward the ovarian region and probably would be a more effective exposure for the production of gene mutations than irradiation from an atom bomb. However, this is a small amount of information and would need confirmation by others.

The paper by Uchida et al [1975] shows that mitotic nondisjunction of human lymphocyte chromosomes can be induced by exposure to a low dose of radiation. This result is from in vitro treatment with x-rays and not in vivo effects of medical treatments. Uchida is convinced that radiation increases nondisjunction, which results in chromosome anomalies among the offspring of irradiated persons. A recent paper by Goad et al [1976] shows the incidence of aneuploidy, usually one extra chromosome, to be not only endemic but also epidemic. There is not only a basic frequency of aneuploidy, which might result in part from radiation, but apparently a nonrandom increase in aneuploidy in babies born in Denver, Colorado, hospitals between the months of May and October each year for several years in succession. Both the constant endemic and the seasonal epidemic rates probably are determined by environmental factors of various kinds which may well include natural radiations. However, we are again dealing in reasonable suppositions rather than established facts.

The reason it is so hard to provide proof of genetic results from irradiations in people is that they are so difficult to work with. We cannot deliberately irradiate human gonads and then test the offspring for recessive lethal mutations as Muller did with his fruit flies. One of the most thoughtful considerations of this problem is included in the position paper by Neel [1971]. He assumes that exposure to radiation and chemical mutagens may be increasing mutation rates and tackles the theoretical problems involved in attempting to evaluate the assumption. This is of practical importance because a potential mutagen could pass screening tests with mice and yet alone, or in combination with other agents, be mutagenic in man.

The only question we will consider here is that of how large a population would be needed to detect a 50% increase in mutation rates for a few traits within one year. Neel shows that even for a drastic increase of 50% in the mutation rate, one would need two samples of 195,911 births (at the 0.01 level of statistical significance) in order to demonstrate this increase in the mutation rate. It would take a population of about 10 million people to produce 195,911 births in one year. This required sample size is larger than any studied to date and helps to answer the question as to why we have little direct evidence about the effects of radiation on the mutation rate. Sample sizes have been much too small to demonstrate increased mutation rates due to any cause in man. Even in mice and swine, which received relatively large doses of radiation over a series of generations, there was no statistically significant demonstration of inherited deleterious effects. But here again the sample sizes were too small and the techniques too inefficient to demonstrate increased mutation rates above those due to natural causes.

No dose of irradiation, no matter how small, can be considered harmless. The fact that it is impossible to prove small doses are harmful is merely a demonstration of technical inadequacy, because theoretically a single ionization should be able to cause a mutation. Every reasonable precaution should be taken to prevent unnecessary irradiation of any kind. On the other hand, some radiation risk must be accepted as a trade-off for the tremendous benefits which radiation provides for medical and other uses. The decisions as to what risks are permissible should be based on the opinions of experts representing all technical areas as well as the general public.

The most difficult problem concerns the possible genetic effects of low level radiation. Both men and women who have been exposed to the common sources of low level radiation ask about the expectations of damage to future children. Usually we don't know what their risk might be. It is presumed to be low so that amniocentesis would not be likely to find chromosome aberrations. However, amniocentesis may be worthwhile in some cases in order to allay hyperanxiety. Obviously the physician should always ask whether a patient is pregnant before exposure to diagnostic radiation. Gonads should be shielded when appropriate.

Grahn [1972], in an excellent brief review of the problems germaine to research on the genetic effects of low level radiation in people, points out that the endpoints of fetal death, sex ratio, malformation rate, and neonatal and infant mortality and morbidity are either intrinsically imprecise, or have too small a genetic component, or both. Present day efforts to demonstrate genetic effects of radiation at fractions of natural background levels must therefore have little genetic or epidemiologic substance.

In Minnesota, there has been a physical confrontation between farmers and surveyors representing the power company which has installed high voltage power lines. The power company has obtained the necessary permits but the farmers claim that the ozone from the power lines will damage their crops. The basic quarrel is between the farmers and the people of Minneapolis and Saint Paul. The cities need the electricity, but the farmers insist that they get along without it, rather than have the power lines pass over their land. The dispute had not been resolved at the time this was written. There is no question but that power lines and nuclear reactors must be established, but at the same time the objectors must be mollified in some honest way. Trade offs must be accepted so damage and benefit are balanced as well as possible.

The wide interest in the possible hazards associated with nuclear power plants has fostered attempts to apply techniques of risk-benefit analysis to other potential hazards, such as food additives and drugs, dams and use of coal, and so on. Eventually such large questions are acted upon by the voters, and in June, 1976, the voters of California refused to prohibit the construction of new nuclear power plants. In considering the potential hazards of a particular system, such as medical radiation, it is important to weigh them against the "other risks," such as drunken driving, in order to maintain a proper perspective and to allocate the "safety" dollar in correct proportions.

It seems clear from the scientific evidence, that, after conservation of energy and more efficient use of it, the choice for new or replacement plants before the year 2000 must be between coal and nuclear power. The greater emphasis should be on nuclear power so that coal can assume its proper long-term role as a chemical resource as well as a power source. Careful reviews of this subject have been provided by Rose et al [1976] and by Cohen [1976]. The latter author writes with an admirable clarity and puts the whole problem into sharp perspective. He points out that cigarette smoking, driving an automobile, working in a coal mine, or even transporting coal are all much more dangerous than living next to a nuclear power plant.

Twenty years ago many of my counseling cases were engendered by fears that genetic damage had been done to the person by medical radiations, radar sets, and other sources of radiation which the public suspected of being hazardous. Apparently the public fears of that period have been assuaged as the queries now are related to the possible genetic damage caused by heroin, LSD, and other addictive drugs. Everything that one does involves a risk; every act is a trade off and should be viewed in that perspective.

It seems reasonable to me that the use of nuclear power must increase in some parts of the world, at least, and that it will be one of the safest forms of energy use. A major nuclear war would be likely to destroy civilization but not humanity as a whole, as there would be some survivors in out-of-the-way places on the globe.

THE DRUG SCENE

Addictive and Non-Addictive Drugs

The physician cannot help but wonder whether the chemically dependent patient is entirely the product of his environment or is there also a genetic component of the dependency? The question asked of the physician these days is a more direct query. It is from a pregnant woman who has used LSD or marijuana and wants to know whether her child is going to be damaged as a result. Or it may have been the father of the expected child who was the drug user. The reason this question is asked so often is that the general public has read various accounts in the popular press of the damage done to chromosomes by the various drugs used so widely by young adults. What "hard" data are available regarding chromosome damage in the drug user and the likelihood of some malformation in the child?

Many of the drug users take numerous different drugs concurrently, which confounds the picture for any particular drug. However, this complexity is not an immediate problem because the reviews of the numerous studies on this subject agree that there is no certain evidence that any of the drugs or combinations of them cause harm to *in vivo* human chromosomes or to the human embryo *in utero*. Large doses of most any drug added to cells *in vitro* would probably damage the chromosomes, but we are not concerned with that. Our interest is only in dosages actually taken by people and not in the usually massive doses administered to animals or cell cultures in the laboratory.

LSD has been given therapeutically as an adjunct to psychotherapy, and in this situation there is presumably control over ingestion of other drugs. Robinson et al [1974] studied the lymphoid cells of 50 such patients and 50 controls for chromosome aberrations. There was no significant difference in chromosome breakage in the patients compared with the controls.

It is hard to believe that LSD, marijuana, tobacco smoking, and other drug taking is good for you, but on the other hand there is no *strong* scientific evidence that the person's chromosomes are damaged or that there is an increase of anomalies among the children of chemically dependent parents. Perhaps such evidence will be obtained in the future, if it is possible to resolve some of the experimental difficulties in obtaining satisfactory data.

We know that great *social* damage results from the use of illicit drugs such as heroin and LSD, particularly among the young. We are inclined to forget that alcoholism is much more damaging to society than the other drugs because it is an addiction of millions of people of almost all ages. No one ever asked me whether heroin addiction has a genetic basis, but the question as to whether alcoholism is hereditary has produced anxiety and confusion for hundreds of years. In the early years of this century most geneticists assumed that chronic alcoholism had a simple genetic basis. It was considered to be related to mental retardation, the psychoses, and social ills in general. The association between these defects of society is as real today as it was then, but the causal factors do not seem to be as clearly related now as they were thought to be in the early 1900s.

Alcohol is the favorite mood-altering drug in almost every human society. Its psychic effects, both pleasant and unpleasant are well known. What is less well known is that it is a toxic drug which damages the liver and can cause disability and death. Lieber [1976] pointed out that alcoholism is the third most frequent cause of death between the ages of 25 and 65 in New York City. Fats and most carbohydrates can be oxidized by most tissues but alcohol must be oxidized in the liver, which is why the damage is caused there. The alcoholic spends much of his time time and energy drinking and tends to disregard such routine tasks as eating. The alcohol contributes directly to the lack of interest in food and damages the digestive organs, also, directly. Death results as a progression from "fatty" liver to alcholic hepatitis to cirrhosis to hepatic coma.

Numerous investigators have demonstrated strain differences in alcohol preference in mice. The heritabilities for these reactions to alcohol seem to be real but not high, in most cases less than 0.20, as reviewed by Whitney et al [1970]. Thus there are genetic differences between strains of mice in the degree to which they prefer alcohol instead of water.

There are well-known human ethnic differences in their reactions to alcohol. American Indians are said to have social problems due to a high rate of alcoholism. For some time there has been a lively interest in the possibility of racial differences in genes having to do with the type of alcohol dehydrogenase present. These enzymes along with aldehyde dehydrogenase are responsible for alcohol oxidation. According to Stamatoyannopoulos et al [1975], human alcohol dehydrogenase is a dimer formed with the random association of three types of subunits (α, β, and γ) encoded at three structural gene loci designated ADH_1, ADH_2, and ADH_3. There is an opportunity in this complexity for gene mutations to have occurred which could be present in different frequencies in different races, and thus cause different reactions of various races to alcohol consumption. Indeed, they found this to be the case. A Japanese sample contained 85% of the persons who were homozygous or heterozygous for an atypical gene, $ADH_2{}^2$, while only 15% were homozygous for the usual $ADH_2{}^1$ gene most frequent in Caucasians. We do not understand the correlations between the genetic differences demonstrated by electrophoresis and the greater susceptibility of "Mongoloids" to alcohol. Wolff [1973] showed that vasomotor flushing of the face in response to small quantities of alcohol significantly differentiates Japanese, Chinese, Koreans, and at least one tribe of American Indians from matched groups of Caucasoids. This suggests only that there are significant genetic differences between people of Asiatic and European derivation in their reactions to alcohol. While the major practical discoveries in this area are still to come, it is clear that in thinking about the problems of alcoholism one must abandon the idea that they are entirely environmental phenomena and accept a small but significant genetic component of the variation between people.

There are many remaining questions of genetic interest such as this one: are the genes related to alcohol preference different from those related to the physiological phenomenon of alcohol dependence? Much work remains to be done, and it would seem that it should be possible to design significant experiments with the techniques now available. Other questions are of great importance, such as the psychiatric query as to whether there is a general psychological state which is especially prone to accept drugs of all kinds in order to obtain mood changes.

Genetic counseling is still chaotic relative to the drug scene as well as other areas.

Non-addictive Chemicals

The general public tends to confuse mutagenic and carcinogenic chemicals, and this is entirely understandable. Indeed, many of the same agents cause both mutations and cancers, at least in laboratory organisms. Miller and Miller [1971] reviewed a vast amount of literature and concluded that there is, "a general qualitative correlation between electrophilic reactivity, carcinogenicity, and mutagenicity wherever enough information is available on the nature of the probable carcinogenic metabolites."

Ames et al [1976] state that, "Very few chemicals in general are mutagens or carcinogens, and the finding that more than 90% of carcinogens tested have been detected as mutagens (and that almost every mutagen that has been given an adequate cancer test is a carcinogen) may actually mean something." Indeed it must mean something and once it is known what the meaning is, substantial progress toward cancer prevention and treatment should be possible.

If carcinogens cause the neoplasm by producing a gene mutation first (one of the popular theories) then carcinogens are mutagens. All that can be stated now is that many, and perhaps all, chemical carcinogens are potential mutagens. Similarly, many, but possibly not all, mutagens are potential carcinogens. Eventually we will have a better idea as to the molecular correlation between carcinogens and mutagens. The general public is rightfully concerned about carcinogenetic chemicals which may be present in foods, bodies of water and in the atmosphere. In Minnesota there has been a celebrated legal battle for some years regarding the discharge of taconite tailings into Lake Superior by the Reserve Mining Company. The tailings include microscopic particles of asbestos which are alleged to cause cancer. There is no direct proof that this naturally occurring asbestos does cause cancer when imbibed with the water, but environmental proponents prefer to close down the plant rather than take a chance on the possible health hazard. The courts have ruled that Reserve Mining must cease discharging the tailings into the lake by July, 1977. The ruling has been appealed and it is astonishing how long it takes to resolve complicated disputes of this type. Some years ago cyclamates were banned because of possible carcinogenic properties. The companies affected lost millions of dollars, via their stockholders, and there is still only evidence from laboratory animals that these sweeteners are carcinogenic. Even saccharine is now to be banned. The cynical may wonder if the FDA will recommend sometime that eating itself be discontinued? None of the above problems have been resolved on a biological basis.

A report of an association of chromosome breakage and birth defects with spray adhesive exposure resulted in a ban on the sale of these products and nationwide publicity warning exposed women. Six months later the ban was removed; the association was not confirmed. Hook and Healy [1976] point out that this false identification of an environmental agent as a mutagen or teratogen resulted in unnecessary consequences such that eight "exposed" women elected to abort their fetuses. More than 1,100 inquiries on the subject were received by medical genetics centers in the United States.

There may be instances where restrictions on the use of some products were not warranted. However, the prohibition of the use of thalidomide in the United States paid off handsomely in preventing the birth of children with phocomelia; one of the heroic stories of governmental control of commercial products. The overriding question of course, is that of determining the balance between risks and benefits to be obtained from the host of new products which flood the markets every day. The complications of the relationships between government and industry are unbelievable and demand levels of honesty and intelligence which are rare, indeed, in order to serve the people best.

The genetic counselor gets frequent questions concerning the possible effects of various chemicals on the yet unborn and is left in the unsatisfactory position of not knowing any of the answers. We do not even have empiric risk figures in this area, and those having the questions resist our efforts to tell them that we really do not have data with which to provide answers. The dilemma is made more acute because the general public would not permit the experiments on people that could provide the answers. Oftentimes political questions involving race and social class get unnecessarily entangled with health problems. The complexities related to public interest and participation in health problems are likely to increase rather than decrease.

There is an enormous number of chemicals which could be mutagenic, so we are concerned about them. In many cases we don't know whether the chemical enters the human gonad or not and if it does, whether gene mutations are produced. It would be foolish to ban the use of all chemicals because we don't have this information.

An amusing but real problem is that related to temperature and mutations. In Drosophila, an increase of $10°C$ increases the mutation rate two- or threefold. A study of Swedish nudists indicated that the scrotal temperature of human males in ordinary clothing is about $3°C$ higher than that of nude males [Ehrenberg et al, 1957]. The higher temperatures due to clothing could increase the mutation rate, which led the investigators to suggest that the wearing of pants might be more dysgenic than the radioactive fallout from the testing of nuclear devices carried on at that time. No practical solution of this problem is known! All of the above shows mankind to be so completely integrated into the environment that it is difficult to separate out the specific environmental factors which must be controlled for our greatest benefit. Clearly there is no free lunch, and both sides of the balance must be carefully evaluated in order that the trade off be best for the largest number of people.

The problem for genetic counseling resulting from the vast array of drugs taken by the ton by the general public are clearly too numerous to be taken up, one by one, in this book. The experiments in which large doses of a drug are given to laboratory animals are necessary but not sufficient to tell us what would happen when small doses of the drug are administered to people. Consequently, when we get a request as to how drug X will affect the fetus of a woman who took the drug during her pregnancy, what can we say? There are no appropriate data, except for a few drugs such as dilantin, and even in these cases the results are "soft" science. My response has been to tell the pregnant counselee that we cannot answer her question specifically, but that as a generality there is probably no great danger. Probably the danger is substantially less in reality than that which she has feared.

THE FETAL ALCOHOL SYNDROME

Apparently, Greek and Roman mythology suggests that maternal alcoholism at the time of conception can lead to serious problems in fetal development. The question of the effect of alcohol upon the embryo was attacked vigorously by many scientists around the turn of this century but with rather equivocal results. Interest in the problem lapsed for some years but recently concern with the fetal alcohol syndrome has become quite fashionable.

In 1973, Jones et al set forth a unique pattern of malformations, which they referred to as the fetal alcohol syndrome in the offspring of eight chronically alcoholic women who continued to drink heavily during their pregnancy. The syndrome is a specific pattern of altered growth, structure, and function which is identifiable at birth. Birth length was more severely affected than birth weight. The children were mentally retarded in most cases. About one in one thousand births seem to display the syndrome. Microcephaly seems to be the most distinctive anomaly. Cardiac anomalies and the other effects of maldevelopment are found in varying degrees as would be expected in a syndrome where the damage is generalized.

There is general agreement that heavy drinking throughout pregnancy is likely to damage the fetus. What the minimal permissible dose for which there would be no fetal damage is, has not been established. There must be some minimal consumption which would not be damaging, certainly a teaspoonful spread over a day would do no harm.

Ouellette et al [1977] showed that 32% of infants born to mothers who drank alcohol heavily during pregnancy had congenital anomalies compared to 14% among their moderate drinkers and 9% in the group of abstainers. The most serious anomaly seemed to be microcephaly, though no alcohol abuse syndrome was identified, other than lighter weight of the affected baby. Heavy smoking also results in lighter weight babies. However, heavy smoking and heavy drinking are also correlated with each other. Kline et al [1977] showed that cigarette smoking in pregnant mothers also almost doubled the risk of a spontaneous abortion.

Ouellette and some others do not consider the damage caused to the fetus by alcohol to be a syndrome in the formal sense of the word. This is a rather academic point. The bottom line is that drinking and/or smoking are no-nos for women who are pregnant or likely to become so. For others, these two addictions can cause serious damage to society and themselves, but these are moral problems which the genetic counselor had best approach only as a compassionate human being.

INFECTIONS

Rubella

Warburton and Fraser [1959] emphasized that the development of a fetus depends on a precise and intricate system of interactions between *two* sets of hereditary factors and *two* environments all acting at the same time on the developing embryo. The mother and the fetus each have their own environment and their own heredity, so interactions between them all must be complicated indeed.

The relationship of congenital anomalies in the newborn to maternal rubella was established by Gregg [1941] after an epidemic of German measles in Australia. A more serious epidemic occurred in the United States during 1964—65, with about 20,000 babies showing various combinations of cataracts, cardiac defects, deafness, mental retardation or microcephaly, which identify the child as having been infected with rubella in utero. Dudgeon [1975] showed that 15.2% of the children from mothers who were infected with German measles during the first 12 weeks of their pregnancies had gross abnormalities, while infections later in pregnancy did not significantly increase the frequency of defects compared with a control population. Early reports gave estimates of 50% or more for abnormalities if the mother had rubella during the first trimester. These retrospective reports instilled such terror that many physicians were prepared to provide therapeutic abortions with few questions asked. While Dudgeon [1975] showed that 15.2% of children produced by mothers infected during the first trimester, when children who were examined after four years of age were included, the frequency of anomalies rose to 43.1%. Fortunately, the cases discovered in the older children were less severe, of course, than those detected at birth. The total damage done by a rubella epidemic is, however, catastrophic and should be prevented.

There are now anti-rubella vaccines available some of which, unfortunately, may rarely have severe side effects, such as arthritis. Programs for vaccinating all school children do not seem practical, and lack of public concern has limited the scope of vaccinations for women who have a reasonable probability of becoming pregnant. Certainly women who have just had rubella should use contraceptives for at least 90 days afterwards. The rubella problem seems to be an important public health question which does not seem to have been addressed in a very effective way (see Stotts [1978]).

Rubella is not the only organism to cause damage to embryos in utero. There are Toxoplasma, Listeria, Cytomegalovirus, and Herpes simplex to be added to the list. There is relatively little known about their relations to chromosome changes or gene mutations.

The damage to the embryo due to rubella results directly from the invasion of the embryonic cells by the virus and the destruction of cells by its reproduction. Do virus infections of any kind cause damage to the chromosomes in the cell without destruction of the cells? More specifically, could viruses damage the chromosomes of human sperms or eggs and thus cause chromosome aberrations in the individual resulting from such a sperm or egg? Once again, we do not have the answer to our question. Indirectly, the answer would seem to be yes. Human cells grown in vitro and infected with measles and other types of viruses show at least three kinds of changes: the first is single breaks; the second, extensive breakage of the chromosomes; and the third, cell fusion and spindle abnormalities. Nichols [1966] and others have provided this kind of in vitro evidence of chromosomal changes due to viral infections. There is still no direct evidence that viruses have a mutagenic effect which is transmitted through human eggs or sperm.

Slow-Acting Viruses

We come now to the extremely broad topic of susceptibility and resistance to the horde of organisms to which mankind is a host, the viruses, bacteria, fungi, and parasites too numerous to mention. There is not only the problem of resistance to specific pathogens but the broader area of immunogenetics, which includes the nonspecific resistance to all infectious diseases. Presumably genetic mechanisms of all possible types will be involved as aspects of the total picture.

The thymus is the earliest recognizable lymphoid organ in the embryo, and its chief role is thought to be the induction of the differentiation of small T-lymphocytes, which then become immunologically competent, recognizing foreign antigens and, together with the bone marrow B-lymphocytes, responding to them. Immunogenetics as it pertains to blood groups and allergies is discussed elsewhere in this book.

The lymphocytes recognize all sorts of foreign cells as well as microorganisms. Fortunately, they do not react to every gene difference in a transplant, for instance, and only certain gene differences play significant roles in the acceptance or rejection of a transplanted kidney or other tissues. The histocompatibility genes involved are numerous and fairly discrete in their action, so that siblings of the patient are the most likely source of compatible donors. Naturally, it is much more convenient to use organs from persons who have just died and have donated their bodies for this purpose; however, they are less likely to have matching genotypes than are some of the patient's "blood" relatives. Except for identical twins, one would never, or hardly ever, get perfect genetic matching because of the extreme genetic diversity even within single families. Immunosuppressive agents assist the transplant by suppressing the patient's lymphocytes, but this makes the patient vulnerable to infections. Hopefully, future research will produce more specific and less dangerous immunosuppressors.

There is evidence that a person can produce antibodies against certain of his own tissues. This is known as *auto-immunity*. Systemic lupus erythematosus, polyarteritis nodosa, myasthenia gravis, Hashimoto's thyroiditis, pernicious anemia, and rheumatoid arthritis are some of the diseases in which auto-antibodies have been identified. There is a great deal more to be learned about the molecular biology of auto-immunity. We do not have good empiric risk figures for counseling for most of the auto-immune diseases.

There is the rare child who does not produce lymphocytes in sufficient numbers to resist infections of the usual kinds. Bruton's disease is an agammaglobulinemia which causes the child to be excessively prone to pyogenic bacterial infections but not to viruses. This failure is due to a sex-linked recessive gene in Bruton's disease. There are perhaps 20 separate disorders due to immunologic deficiency. In every case there is the all important environmental factor, the infecting organism.

The slow-acting viruses are among the most interesting of the various groups of environmental agents. Some years ago the rare neurological disease called Kuru was found in the Fore tribe of people in New Guinea and thought to be the result of simple Mendelian genetics. However, Gajdusek et al [1967] found that they could transmit the disease to chimpanzees. It was then found that the causative agent was a virus-sized organism, similar to that causing scrapie in sheep, with an exceedingly long incubation period occupying a large fraction of a life time. Creutzfeld-Jacob and a few other rare neurological disorders behave in a similar fashion. These diseases are too rare to warrant more space but they stimulated renewed interest in a more frequent neurological disorder, multiple sclerosis.

Multiple Sclerosis

Multiple sclerosis is more prevalent in northern climates (over 60/100,000) than in southern ones (less than 10/100,000). There is a familial association of cases, but the concordance of identical twins is so low that the genetic basis must be rather weak. Yet it is not highly contagious, and it is not associated with socio-economic status. Myrianthopoulos [1970] provided a comprehensive review of the epidemiologic and genetic data. Henle et al [1975] have shown that there is a multiple-sclerosis-associated agent in the brain tissue and sera of patients which can be serially transmitted from mouse to mouse by passage of brain homogenates.

The evidence from the Henles, Koldovskys, and associates is that there is a new microorganism strikingly associated with multiple sclerosis. It has not yet been seen in the electron microscope. The research is still tentative, but if there is a reproducing viral agent which is causative of multiple sclerosis, then it should be possible to produce a vaccine and thus prevent many cases of the disease. However, the picture becomes more complicated all the time. It has been known for a long time, Adams and Imagawa [1962], that the measles antibody titers of patients with multiple sclerosis tend to be higher than those of their control populations. Paty et al [1976] and others have shown that specific HL-A types also have higher frequencies in multiple sclerosis patients than in control populations. They confirm a familial factor in elevated measles antibody titers and suggest that HL-A antigens are linked to one of the factors that determines measles antibody titers in multiple sclerosis patients. They predict that a better understanding of the constitutional factors in susceptibility to multiple sclerosis will be able to shed light on why some patients run an exceedingly benign course while others progress very rapidly to death within a few years.

Alter et al [1976] searched 10 families for evidence of genetic linkage between a possible gene for multiple sclerosis (1 abeled MSS) and the HLA gene complex. The association approached but did not reach statistical significance. A study of more families is warranted.

Multiple sclerosis has received such wide interest lately that three issues of *Science* contained a continued series of articles about it written by Maugh [1977]. The reader is referred there for a splendid review of this exceptionally complicated disease, which results from a complex interaction of genetics, environment, geography, viruses, and the patient's own immune system. The only redeeming feature of such a complicated system is that the empiric risk figures will be low. The risk for first degree relatives is probably in the vicinity of 2%–3%.

21
The Rare Genetic Traits

Most physicians have little interest in the rare genetic diseases such as ectrodac-
tyly, Thomsen's myotonia, retinoblastoma, ichthyosis hystrix gravior, and so on.
Even the hospital pediatrician will only see a small percentage of the total mass of
rare mendelian genetic traits, most of which are listed in McKusick's [1978] excellent catalog. If the physician is never going to see most of these traits, why
should he be interested in any of them? He will see the frequent central nervous
system defects, the Down syndrome child and the like, but why spend valuable
time on the rare and often obscure defects, even if they do obey the simple
mendelian laws?

It is true that no one person can become a specialist for all of the rare genetic
traits. There is no way in which the family physician can remember all of the
catalog of traits. However, he can expect to encounter a few of them in his practice and there are a few general considerations which will apply to all. The probability that any specific rare trait will appear in his practice is very small, but the
likelihood of some one or more of the rare traits showing up is substantial. There
is also a high probability, when the rare trait does appear, that heroic surgery or
elegant metabolic studies may be required which will entail referral to some large
medical center. A diagnostic work-up and long-term treatment there may be
necessary. Presumably genetic counseling will be offered at the center, but it is the
responsibility of the primary physician to check with the parents to ensure that
it has been provided and perhaps add some helpful thoughts of his own.

Genetic counseling is particularly valuable for the rare traits because frequently it may be statistically precise. Relevant techniques such as amniocentesis may
work very well for some of the rare traits like Tay-Sach's disease and other enzyme
defects and chromosomal anomalies. The rare traits as a group count for a substantial part of a genetic counselor's load, so we should see how he goes about
providing this service for the major types of rare mendelian traits. One reason that
these traits are rare is that they are so serious that the affected child dies or seldom reproduces even when adulthood is achieved. The supply of such genes is
therefore continuously reduced in the population and must be replenished by

mutations, which are rare phenomena and determine the frequency of the trait either directly or indirectly.

The relationship between the mutation rate and the rare dominants is easy to understand. The birth of a child with a regular dominant anomaly from unaffected parents is good evidence that a new mutation for the trait occurred in one or the other of the parents and was transmitted to the child. The child, if it reproduces, would expect half of its offspring to be affected, because all of the cells of the child-become-parent would have one member of the chromosome pair in each cell with the newly mutated gene present. What are the counseling risks for the unaffected couple who produced the affected child? Hopefully, the new mutation occurred only in a small sector of one ovary or one testis of the unaffected parent and, if that is the case, the chance of another affected child would be so small that the parents could ignore it. On the other hand, it is possible that the mutation occurred early enough in the embryogeny of the gonad so that there would be many descendant cells with it, and the chances of another affected child would be increased. The risk might even approach 50%. We face a very uncomfortable dilemma, as the most probable risk is under 1% and if so, can be ignored, but there is always the possibility that it would be higher than that. At this point one must consider the burden of the trait involved. The parents can weigh this burden to some degree, as they have already produced their first affected child. If the situation is extremely stressful, the couple may decide to refrain from having any more children and request the physician's written support or help in obtaining a sterilizing procedure. If the burden is not too heavy, the couple will probably decide to go ahead with another pregnancy — especially if they have no other normal children.

The most significant genetic counseling cases are for congenital traits, and probably the obstetrician will provide this service. Traits not present at birth, but which appear in infancy, would fall into the province of the pediatrician, while those with a later age of onset will be seen by the internist, neurologist, or usually first by the family physician.

The late-onset neurological disorders which have the mendelian dominant type of inheritance are particularly insidious because the person often has completed reproduction before the onset of the disease. Huntington's disease, Marie's cerebellar ataxia, and such disorders receive more than their ordinary share of attention because they are such frightful afflictions and the genetic counseling is difficult. The counseling situation often involves young couples in which a parent of one of the partners has Huntington's disease. The young person wants to have children, but does not know whether he is free of the gene or whether his 50% chance of carrying it has gone against him. The older he gets without showing any symptoms of the disease the smaller the chance that he (or she) is carrying the gene for it. The average age of onset for Huntington's disease is about 44 years, which means that half of the eventual cases will have been diagnosed before that age, and the other half afterwards. Patients in our Minnesota study have been diagnosed between the ages of 18 months and 74 years. We were inclined to consider the 74-year-old man free of the Huntington's gene until his diagnosis was made. Until then his chance of carrying the gene for Huntington's disease, at say age 70, might have been about 5%, while at his birth it was 50%. When his symptoms were observed at age 74 his chances of being a carrier of the gene changed dramatically to 100%.

It is conceivable that a test will be found which will detect the carrier of the gene for dominant, late-onset disorders such as Huntington's disease, but we will then have the ethical dilemma as to whether it is appropriate to tell the carrier of his likely death from the disease. If there were a "cure" for the disease, then a screening test used to inform the carriers (and the 50% who do not carry the gene) would be very helpful. The disease could be almost eliminated in one generation if a reliable test were available and the carriers of the gene refrained from reproduction. Many carriers might be willing to accept a sterilizing operation, as they are well aware of the great tragedy which accompanies Huntington's disease, and they would not wish to take any chance of passing it on to their descendants. Shokeir [1975] has discussed the biochemical rationale of screening tests for Huntington's disease.

The rare autosomal recessive diseases present genetic counseling problems very different from the dominant traits. Quite often, perhaps a fifth of the time, it will be found upon questioning, that the parents of the child with the *rare* recessive defect are blood relatives. Consanguinity contributes to the probability of homozygosity for all genes, and any rare recessive gene has a much greater chance of becoming homozygous in the offspring of blood relatives than in the children of unrelated parents. The reasons for this have been explored in an earlier chapter of this book. It is quite probable that the consanguineous parents will have strong feelings of guilt, as the general public is well aware of some vague relationship between the consanguinity of parents and defects in their offspring. The parents no doubt will be aware of the general attitude and may have had genetic counseling before their marriage. Just how such guilt feelings could be relieved is not clear to me, other than by general sympathetic attention to the couple. A psychiatrist might be of assistance in removing the uneasiness related to the incest taboo which probably extends to more distant relationships to some degree.

Most couples who have come to the Dight Institute for genetic counseling have an ambivalent view about the 25% risk for subsequent children where a mendelian recessive trait is involved. All couples will agree that a 50% risk is "high," and most consider a 5%–10% risk is "low," depending, of course, upon the expected burden. With the 25% risk there is no firm opinion as to whether it is high or low for most couples. However, for most of the recessive traits which result in genetic counseling the burden of the trait is quite heavy. Albinism, for instance, is considered to be a moderately heavy burden. Recessive types of severe mental retardation are considered to be very heavy burdens. Mental retardations with severe physical stigmata such as the Sjögren-Larsson syndrome are especially heavy burdens. Educated parents are not anxious to accept any risk of a repetition of such serious recessive traits, and for these the 25% chance is too high to take. One of the major missions for medical genetics of the future is to decrease the number of people uneducated in regard to medical genetics, because their ignorance will not be bliss after children with genetic defects are born.

An important characteristic of recessive traits is that they are usually due to a deficient or defective enzyme. The heterozygote can be detected in some cases because of the partial deficiency of the relevant enzyme. This permits population screening for the heterozygotes and prenatal detection of the homozygotes for some recessive traits like Tay-Sachs disease. Therapeutic abortion of the affected fetus is possible in such cases, and the abortus can be replaced with a prenatally tested unaffected child.

Parents of a child with a rare recessive trait almost always point out to me in great wonderment that there has never been a person with the trait among their ancestors. This is correct information in most cases and not surprisingly so. The rare recessive gene may be passed down through the generations for hundreds of years without having become homozygous. Individual dominant and sex-linked deleterious genes are likely to be eliminated in a few generations because of their effects on the bearer. Recessive genes in the heterozygous condition have little effect on the carrier, by definition, and their survival time in generations would depend largely upon genetic drift, a sort of random walk. However, many recessive genes are probably not completely recessive; consequently, natural selection would tend to eliminate them faster than completely recessive genes.

Recessive genes located on the sex chromosome are in a different category from those on one of the other sets of chromosomes. A single sex-linked gene in a male is expressed because there is no corresponding gene to provide enzyme, as is the case in ordinary heterozygotes. The term "hemizygote" is used in the male to emphasize that the X and Y chromosomes do not represent homologous pairs as do the autosomes. An X-linked gene becomes manifest in the male because there is no "protection" from any homologous locus on the Y chromosome, as there would be for an autosomal recessive. If a trait is transmitted from father to son it is *not* sex-linked.

The transmission of X-linked dominant diseases is from the affected male to all of his daughters and none of his sons. The heterozygous female is always affected in X-linked dominant diseases, but her disease is generally milder than that in the affected male. Among her offspring, half would be affected, irrespective of their sex.

The differences in transmission of sex-linked traits compared with autosomal traits permit more precise counseling in some cases. For instance, if a male has a dominant X-linked trait one knows that all of his daughters, but none of his sons will be affected. One could abort all daughters, after looking at the chromosomes, and save all sons. The situation for recessive X-linked traits is not so precise. Here the female carrier expects half of her sons to be affected but none of her daughters to show the trait. If amniocentesis and chromosome study were requested, then all daughters would proceed to term, but all sons would be aborted – the latter poses an ethical dilemma because half of the aborted sons would be normal. Hopefully, prenatal diagnosis will become available for all severe X-linked recessives so that the 50% of the males that are normal can proceed to term.

The sex-linked recessives illustrate the relationship between risk and burden very well. *The risks* for women carrying the gene for red-green color blindness or for Duchenne type muscular dystrophy are the same; 50% of the sons will be affected. There is no cure for either anomaly, but the *burden* of color blindness is so trivial it can be ignored, whereas the *burden* of Duchenne muscular dystrophy is catastrophic – it not only results in the death of the boy, but seriously disrupts relationships in the whole family. There is an old saying for genetic diseases, that where there is a sick person there is a sick family. Those of us who have suffered along with these families know that the saying has much truth in it.

We have not supplied a chapter on deafness or blindness, which are general categories similar to mental retardation. However, with deafness and blindness we are able to identify specific diagnoses more frequently than is the case with mental retardation, but most deafness and blindness can be looked at as heterogeneous groupings of separable rare diseases. In many cases the specific diseases have a simple mendelian mode of inheritance or an obvious environmental cause, such as the case of retrolental fibroplasia. There is no way in which we can include a description of each of the defects of the ear or eye in this book. The reader can obtain a splendid overview of the relationship between the various kinds of deafness in the book by Konigsmark and Gorlin [1976] on "Genetic and Metabolic Deafness" and realize how difficult it is to catalog the great mass of rare genetic diseases.

We are extremely fortunate to have the 975-page computer print-out catalog of the rare mendelian traits provided by McKusick [1978], which provides descriptions and references for the several thousand genetic diseases now identified. A glance at the McKusick catalog should impress everyone with the vast array of diseases which have a specific genetic cause.

There is a most important cautionary note which must be heeded by the genetic counselor. It is the problem of the heterogeneity of genotypes for the same phenotype. Leber's optic atrophy, as an example, is inherited as a mendelian dominant trait in some families, as a sex-linked recessive trait in other families, and as an ordinary autosomal recessive in still other families. The list of anomalies which display more than one type of mendelian inheritance gets longer each year. The risk figures for the different types of heredity are significantly different, so the counselor must make every effort to determine which sort of mendelian mechanism is involved in the family concerned. It may be possible to rule out sex-linked recessive inheritance, if the affected person is a female. However, sometimes there will be no way of excluding all but one pattern of inheritance and the counselees must be instructed accordingly. We have considered the problem of heterogeneity in Chapter 4.

Statistically speaking, a risk of 50% or more is high, 25% is medium, and below 10% is low, but it is the burden or seriousness of the trait which is the major concern of the counselee. The 50% risk of the inheritance of campodactyly, a dominant trait causing crooked little fingers, is not a cause for alarm while the 5% risk of the repetition of a neural tube defect is very threatening; though couples who have no normal children are usually willing to take this risk in order to have a normal child.

The final summation is that, though the rare genetic traits are individually infrequent and often result in a heavy burden to all concerned, it is the sum total of all of them that is of great significance and amounts to an enormous health problem. Consequently, correct genetic counseling for each rare trait is of the greatest importance.

22
Putting the Puzzle Together

The summing up or pulling together of our material on genetic counseling should perhaps start with a caveat or disclaimer that genetic counseling is not synonymous with human or medical genetics. One need only thumb through the volume on "Genetics, Law, and Social Policy" by Reilly [1977] to see that genetic counseling is only one segment of activity in human genetics. Genetic counseling is interesting to the public, but it is not a spectacular service and the practice is very private and not open to public view, in individual cases.

The physician who does genetic counseling has a problem which the non-physician counselor is free from. The physician counselor is likely to confuse his roles as a medical man with his role as a geneticist. The two should be kept separate; for example, the physician who is caring for a newborn child with a genetic anomaly may have to decide whether to "pull the plug" and let the child die rather than continue heroic efforts to postpone an inevitable death. This is a legal-moral problem, but it has nothing to do with the genetic counseling which the physician might provide for the parents regarding a subsequent child. The public probably would fail to see that the two responsibilities are independent of each other; hopefully, the physician can see the difference between the broad area of medical genetics and its distinct subdivision of genetic counseling.

The physician who is competent to make the correct diagnosis of a trait has a good chance of being unprepared to do acceptable genetic counseling. The reasons for this are only too obvious when one considers our medical education, as it is today and has been in the past. At the University of Minnesota only an insignificant percentage of the graduates of the medical school have had an *adequate* training in human genetics, have ever attended genetics conferences, or have had a working knowledge of genetic counseling. Perhaps one or two percent of our graduates and less than 10% of our residents have had a modest training in the techniques of genetic counseling. It would be absurd to assume that every physician is competent to diagnose everything and is willing and able to do genetic counseling as well. Only a few qualify.

The PhD geneticist who does genetic counseling knows that he is not permitted to make diagnoses and must obtain them elsewhere. Only a small percentage of genetics PhDs are able or willing to do genetic counseling. Only a few qualify for genetic counseling. Obviously, numerous persons with different backgrounds should be involved in the process of genetic counseling. The counseling should be done by the person who is best prepared to do it in his or her geographic location. Success in counseling depends to such a great degree upon personal characteristics and ability to teach and communicate that any argument as to whether some specific academic degree guarantees successful counseling is non-productive.

If the genetic counselor does not happen to have an MD, he or she will have to learn much of what the physician has already learned. Not the name of every muscle, as most physicians do not know the name of every muscle very long after their anatomy course is finished. Latin and Greek roots are helpful and some course work in pathology, neurology, and pediatrics should be included in obtaining a general understanding of the medical scene. The learning process should continue until death or retirement from intellectual experiences has occurred.

It was shown earlier in the book that the potential demand for genetic counseling should be much greater than the potential supply of counselors. Probably every family has some question about their genetics which it would be helpful to them to have answered. If there is a demand for counseling from millions, then thousands of counselors will be needed; it is not clear to me where these thousands of counselors will come from. It is unlikely that large numbers of physicians will be available for this task. Nevertheless, the benefits to some families from genetic counseling will be very great, and it would seem that the cost-benefit ratio would justify many more genetic counselors than there are now. This should be true, particularly on a worldwide basis.

The major uncertainty about genetic counseling in regard to future demand seems to be a problem of economics. How is the service to be supported?

The family physician cannot afford to do much genetic counseling, economically speaking. In my genetic counseling it takes about an hour of my time to listen to and talk with the person or persons involved. There may be much more time spent in obtaining tests and other technical information and in other contacts over the years. However, my own family physician does not spend an hour at one time with me. Someone else weighs me, takes a blood sample, and carries out various routine procedures preliminary to my brief contact with my physician, as he hurries on to the next waiting patient. Genetic counseling would not be successful in such a "mechanized" health care system, efficient as the system may be for some purposes. I am not complaining about my family physician, I am merely pointing out that his mode of operation is very different from what is required for successful genetic counseling.

It is absolutely essential that the medical diagnosis needed for genetic counseling be as competent as possible, if the counseling is related to a medical problem. Once the diagnosis has been established, the genetic counseling begins and it is not very closely related to the kind of experience that the physician has had. It is much more a social-work type of procedure, where the counselor listens patiently to the counselee and visits in an informal way, often approaching the problem in an indirect and certainly a non-authoritarian fashion. I am astonished that physicians who should know better tell counselees flatly, and without amplification, that they should not have any children. If this is genetic counseling, it is obviously unsuccessful and no one has been helped. Those so treated will turn elsewhere for a less authoritarian and more sympathetic consideration of their problems.

Genetic counseling applies the knowledge gained in the area of human genetics. How much more should the genetic counselor provide than the genetic information required by the definition of the discipline? Should the counselor provide the diagnosis, including specialized biochemistry or chromosome study? Should the counselor provide the psychometrics which may be necessary and the psychological support which may be needed by the parents of an affected child? Should the counselor carry out paternity testing? No one counselor could do all of these things and will need help for most of them, that is, have them done by others.

The counselor must understand that parents always have great expectations for each of their children. They will go to great lengths to avoid having a defective child, because a less than perfect child is a shattering blow to their egos. As Mac Intyre [1973] pointed out

> "It is not unnatural for the parents of a malformed child to experience feelings of revulsion or even hate for that child because of the terrible blow to their egos that its birth has produced. Furthermore, it is not unnatural for such parents to harbor the secret wish that the child would die."

Fortunately, most malformations are not sufficiently grave to cause such an extreme reaction as the one above. Nonetheless, couples do want to give birth to children who are satisfying to them according to their norms. People are willing to accept for adoption children with disadvantages as a charitable response to values they learned at some time in their lives.

It is the psychological problems of the counselees, in addition to the requests from them for genetic education, that give the genetic counselor the most challenging moments of the counseling session or sessions. It is then that the full measure of empathy and support for the counselees is needed. It is difficult for the physician to relax with the counselees because they are even more concerned than the physician about the other people waiting their turn in the waiting room with its ancient news magazines. The family physician is the logical person to do the mass of genetic counseling, but one of the pieces of our puzzle that does not fit in its place very well is the environment of his office, which is usually not very conducive to the deliberate and delicate conversation which is the foundation of proper genetic counseling. I suspect that most physicians are well aware of this difficulty and are glad to refer their counselees to a genetic counseling center, if one is available.

Where are genetic counseling centers available? Usually, they are at universities or large hospitals. The hospitals with genetic counseling centers often have university connections. The relationship between genetic counseling centers and universities is not accidental. Probably university personnel are the only members of the community who are both qualified to do the counseling and have salaries and time schedules which make it possible. The actual locations of the major, and some minor, counseling centers, worldwide, are listed in the National Foundation – March of Dimes International Directory, fifth edition, compiled by Lynch et al [1977].

The state of Minnesota is unique in that it has a clearly identified human genetics unit in the State Board of Health; the unit having been established by the legislature at the request of the Minnesota Human Genetics League in 1959. The director of the Board of Health Human Genetics Unit is Dr. Lee Schacht, who has done a vast amount of genetic counseling there with both cytogenetics and biochemical facilities available. He has also established, along with Robert Desnick, PhD, MD, a modest "satellite" system for genetic counseling which covers much of the state of Minnesota. Richard King, MD, PhD, is now contributing greatly to the satellite system.

One of the pieces of the puzzle which still seems to be missing, and is puzzling to me, is the fact that none of the other 49 states of the union has a human genetics organization of lay people or a legislatively established human genetics unit in the state board of health. Some other state boards of health do have vigorous human geneticists who do genetic counseling while attached to birth defect centers or other divisions of the state health structure. They would have higher profiles, if they had legislatively recognized status. An organization of lay people is extremely helpful in obtaining necessary public recognition and support.

There are also regional genetics centers supported by the federal government and by the National Foundation – March of Dimes and other foundations. The financial support for these is likely to be transitory with uncertain survival for some of them. Even a small endowment is of the greatest value in order to ensure continuity for the center.

Local, state, and federal public health agencies continue to show intense interest in diseases such as rabies, which is now almost extinct, while seeming to pay minimal attention to genetic diseases which can be found in most families, and each of which is more frequent than rabies. It should also be realized that genetic consultation is of value in public health for the planning and execution of programs in such areas as chronic diesease, mental retardation, mental health, congenital malformations, and dentistry, to name a few areas of great importance.

Genetic counseling can be of value not only on an individual basis but within the total framework of public health with respect to control and prevention of disease. As medical advances continue to save more genetic anomalies which were formerly lethal, public health must consider the overall effect of these individuals on the medical, social, and economic structure of the community. Schools of public health and other health training institutions should provide genetic counseling material and techniques for their students, thus providing a background of understanding on which the applications of genetics to public health can be based.

It has been my experience that genetic counseling is overall a pleasant task; a chance to educate, to allay anxiety, reduce fear, and reassure. The genetic counselor generally has the statistics in his favor. Even when the risk and the burden are high, the situation is usually not as bad as the counselees had imagined it to be. There is an important future ahead for genetic counseling. It will be a healthy science to whatever extent it is supported by the health sciences.

Literature Cited

Adams JM, DT Imagawa (1962): Measles antibodies in multiple sclerosis. Proc Soc Exp Biol Med 111:562–566.

Adams MS, JV Neel (1967): Children of Incest. Pediatrics 40:55–62.

Abrams R, MA Taylor, P Gaztanaga (1974): Manic-depressive illness and paranoid schizophrenia. Arch Gen Psychiatry 31:640–642.

Adelstein P, J Fedrick (1976): Pyloric stenosis in the Oxford record linkage study area. J Med Genet 13:439–448.

Adinolfi A, M Adinolfi, MH Lessof (1975): Alpha-feto-protein during development and in disease. J Med Genet 12:138–151.

Allison AC (1954): Protection afforded by sickle-cell trait against subtertion malarial infection. Br Med J 1:290–294.

Alter M, M Harshe, VE Anderson, L Emme, EJ Yunis (1976): Genetic association of multiple sclerosis and HLA determinants. Neurology 26:31–36.

Ames BN, J McCann, C Sawyer (1976): Mutagens and carcinogens. Science 194:132–133.

Ampola MG, MJ Mahoney, E Nakamura, K Tanaka (1975): Prenatal therapy of a patient with vitamin-B-responsive methylmalonic acidemia. N Engl J Med 293:313–317.

Ananthakrishnan R, H Walter, N Hoede, B Morsches, H Holzmann (1974): Enzyme polymorphisms of erythrocytes in psoriasis. Hum Hered 24:53–58.

Anderson DE (1972): A genetic study of human breast cancer. J Natl Cancer Inst 48:1029–1034.

Anderson DE (1975): Persons at high risk of cancer. "Familial." JF Fraumeni, Jr (ed), New York: Academic Press.

Andersen DH, R Hodges (1946): Celiac syndrome V. Genetics of cystic fibrosis of the pancreas with a consideration of etiology. Am J Dis Child 72:62–80.

Anderson RC (1976): Cardiac defects in children of mothers receiving anticonvulsant therapy during pregnancy. J Pediatr 89(2):318–319.

Anderson RC (1976): Fetal and infant death, twinning, and cardiac malformations in families of 2000 children with and 500 without cardiac defects. Am J Cardiol 38:218–224.

Anderson RC (1977): Congenital cardiac malformations in 109 sets of twins and triplets. Am J Cardiol 39:1045–1050.

Annegers JF, WA Hauser LR Elveback VE Anderson, LT Kurland (1976): Seizure disorders in offspring of parents with a history of seizures – a maternal-paternal difference? Epilepsia 17:1–9.

Antley RM, LC Hartlage (1976): Psychological responses to genetic counseling for Down's syndrome. Clin Genet 9:257–265.

Bach FH, JJ van Rood (1976): The major histocompatibility complex – genetics and biology. N Engl J Med, 295:806, 872, 927.

Bear JC (1976): A genetic study of facial clefting in Northern England. Clin Genet 9:277–284.

Beckman L, R Brönnestem, B Cedergren, S Lidén (1974): HL-A antigens, blood groups, serum groups and red cell enzyme types in psoriasis. Hum Hered 24:496–506.

Beckman L, R Lemperg, M Nordström (1977): Congenital dislocation of the hip joint in Northern Sweden. Clin Genet 11:151–153.

Bell AG, MH Cripps (1974): Familial aneuploidy: What risk to sibs? Can J Genet Cytol 16:113–119.

Berkov B, J Sklar (1975): Methodological options in measuring illegitimacy and the difference they make. Soc Biol 22:356–371.

Bertelsen A, B Harvald, M Hauge (1977): A Danish twin study of manic-depressive disorders. Br J Psychiatry 130:330–351.

Bias WA (1975): Exclusion of paternity. Am J Hum Genet 27:245–246.

Bodmer WF (1978): The HLA System Br Med Bull 34:213–309.

Bodmer WF, JG Bodmer (1978): Evolution and function of the HLA system. Br Med Bull 34:309–316.

Bonaiti-Pellié C, C Smith (1974): Risk tables for genetic counselling in some common congenital malformations. J Med Genet 11:374–377.

Borgaonkar DS (1977): "Chromosomal variation in man: A catalog of chromosomal variants and anomalies." Second edition. New York: Alan R Liss, pp 1–403.

Borgaonkar DS, SA Shah (1974): The XYY chromosome male – or syndrome? In Steinberg and Bearn (eds): Prog Med Gen 10:135–222.

Bowman BH, JA Mangos (1976): Current concepts in genetics. Cystic fibrosis. N Engl J Med 294:937–938.

Bowman BH, BJ Lankford, MC McNeely, SD Carson, DR Barnett, K Berg (1977): Cystic Fibrosis: Studies with the oyster cilia assay. Clin Genet 12:333–343.

Boyle IR, PA di Sant'Agnese, S Sack, F Millican, LL Kulczycki (1976): Emotional adjustment of adolescents and young adults with cystic fibrosis. J Pediatr 88:318–326.

Briard ML, J Kaplan, J Frézal (1977): Le conseil génétique III. J Génét Hum 25:77–94.

Brock DJH (1975): Amniotic fluid alpha$_2$-macroglobulin and the antenatal diagnosis of spina bifida and anencephaly. Clin Genet 8:297–301.

Brown MS, JL Goldstein (1976): Understanding the manifestations of receptor deficiency states. Prog Med Genet 1:103–119.

Bulmer MG (1970): "The Biology of Twinning in Man." Oxford: Clarendon Press;

Burdick AB (1977): Frequency of the gene for cystic fibrosis with a view of replacement. and recognition effects and reproduction by homozygotes. Hum Hered 27:366–371.

Carter CO (1965): The inheritance of common congenital malformations. Progress in Medical Genetics IV:59–84.

Carter CO, K Evans (1969): Inheritance of congenital pyloric stenosis. J Med Genet 6:233–254.

Carter CO, K Evans (1973): Spina bifida and anencephalus in Greater London. J Med Genet 10:209–234.

Carter CO, KA Evans (1973): Children of adult survivors with spina bifida cystica. Lancet ii:924–926.

Carter CO, TJ Fairbank (1974): The Genetics of Locomotor Disorders. London: Oxford University Press, pp 1–170.

Carter CO, JAF Roberts (1967): The risk of recurrence after two children with central nervous system malformations in South Wales. Lancet i:306–308.

Carter CO, JA Wilkinson (1964): Genetic and environmental factors in the etiology of congenital dislocation of the hip. Clin Orthopaed 33:119–128.

Carter CO, KA Evans, K Till (1976): Spinal dysraphism: genetic relation to neural tube malformations. J Med Genet 13:343–350.

Caspersson T, G Limakka, L Zech (1971): The 24 fluorescence patterns of the human metaphase chromosome. Hereditas 67:89–102.

Červenka J, RJ Gorlin, VE Anderson (1967): The syndrome of pits of the lower lip and cleft lip and/or palate. Genetic considerations. Am J Hum Genet 19:416–432.

Červenka J, BL Shapiro (1970): Cleft uvula in Chippewa Indians. Hum Biol 42:47–52.

Chakraborty R, M Shaw, WJ Schull (1974): Exclusion of paternity: the current state of the art. Am J Hum Genet 26:477–488.

Ching GHS, CS Chung, RW Nemechek (1969): Genetic and epidemiological studies of clubfoot in Hawaii. Am J Hum Genet 21:556–580.

Christensen AL, J Nielsen (1973): Psychological studies of ten patients with the XYY syndrome. Br J Psychiatry 123:219–221.

Christy M, A Green, B Christau, H Kromann, J Nerup, P Platz, M Thomsen, LP Ryder,, A Svejgaard (1979): Studies of the HLA system and insulin-dependent diabetes mellitus. Diabetes Care 2:209–214.

Chung CS, NC Myrianthopoulos (1975): Factors affecting risks of congenital malformations. II. Effect of maternal diabetes. Birth Defects: Original Article Series. The National Foundation – March of Dimes. New York: Stratton Intercontinental Medical Book Corporation, 11:23–37.

Chung CS, GHS Ching, NE Morton (1974): A genetic study of cleft lip and palate in Hawaii. Am J Hum Genet 26:162–188.

Clark JA, BL Mallet (1963): A follow-up study of schizophrenia and depression in young adults. Br J Psychiatry 109:491–499.

Clarke CA, WK Cowan, JW Edwards, AW Howel-Evans, RB McConnell, JC Woodrow, PM Sheppard (1955): The relationship of the blood groups to duodenal and gastric ulceration. Br Med J ii:643–646.

Clarke CA, JW Edwards, DRW Haddock AW Howel-Evans, RB McConnell, PM Sheppard (1956): ABO blood groups and secretor character in duodenal ulcer. Population and sibship studies. Br Med J ii:725–731.

Cohen BH, JE Sayre (1968): Further observations on the relationship of maternal ABO and Rh types to fetal death. Am J Hum Genet 20:310–360.

Cohen BL (1976): Impacts of the nuclear energy industry on human health and safety. Am Sci 64:550–559.

Conneally PM, AD Merritt, PL Yu (1973): Cystic fibrosis: population genetics. Tex Rep Biol Med 31:639–650.

Cox DW (1964): An investigation of possible genetic damage in the offspring of women receiving multiple diagnostic pelvic x-rays. Am J Hum Genet 16:214–230.

Crandall BF, MAB Brazier (1978): "Prevention of Neural Tube Defects." New York: Academic Press, pp 1–261.

Crandall BF, TB Lebherz, L Rubinstein, WF Sample, J Howard (1979): Outcome of 2,000 2nd trimester amniocenteses. Am J Hum Genet 31:92A.

Curtis E, FC Fraser, D Warburton (1961): Congenital cleft lip and palate: risk figures for counseling. Am J Dis Child 102:853–857.

Czeizel A, C Révész (1970): Major malformations of the central nervous system in Hungary. Br J Prev Soc Med 24:205–222.

Czeizel A, G Tusnady (1971): An epidemiologic study of cleft lip with or without cleft palate and posterior cleft palate in Hungary. Hum Hered 21:17–38.

Czeizel A, G Tusnady (1972): A family study on cleft lip with or without cleft palate and posterior cleft palate in Hungary. Hum Hered 22:405–416.

Czeizel A, T Vizkelety, J Szentpéteri (1972): Congenital dislocation of the hip in Budapest, Hungary. Br J Prev Soc Med 26:15–22.

Czeizel A, J Szentpétery, G Tusnády, T Vizekelety (1975): Two family studies on congenital dislocation of the hip after early orthopaedic screening in Hungary. J Med Genet 12:125–130.

Danes BS, ME Hodson, J Batten (1977): Cystic fibrosis: evidence for a genetic compound from a family study in cell culture. Clin Genet 11:83–90.

Danes BS, B Beck, EW Flensborg (1978): Cystic fibrosis: Cell culture classes in a Danish population. Clin Genet 13:327–334.

Danks DM, J Allan, CM Anderson (1965): A genetic study of fibrocystic disease of the pancreas. Ann Hum Genet Lond 28:323–356.

Dansky L, E Andermann, F Andermann, A Sherwin (1975): Major congenital malformations in the offspring of epileptic women. Abstract, 27th Annual Meeting of the American Society of Human Genetics, p 31A.

Darlow JM, C Smith, LJP Duncan (1973): A statistical and genetical study of diabetes. III. Empiric risks to relatives. Ann Hum Genet (Lond) 37:157–174.

Davenport DB (1923): Body build and its inheritance. Publication No. 329, Carnegie Institute of Washington.

Davenport CB, FH Danielson (1913): Skin color in Negro-White crosses. Publication No. 188, Carnegie Institution of Washington.

Degnbol B, A Green (1978): Diabetes mellitus among first- and second-degree relatives of early onset diabetes. Ann Hum Genet (Lond)42:25–35.

deGrouchy J, C Turleau (1977): "Clinical Atlas of Human Chromosomes." New York: John Wiley and Sons, pp 319.

Dewey WJ, I Barrai, NE Morton, MP Mi (1965): Recessive genes in severe mental defect. Am J Hum Genet 17:237–246.

Dodge JA (1974): Proceedings: maternal factor in infantile hypertrophic pyloric stenosis. Arch Dis Child 49:825.

Doeblin TD, K Evans GB Ingall, K Dowling, ME Chilcote, W Elsea, RM Bannerman (1969): Diabetes and hyperglycemia in Seneca Indians. Hum Hered 19:613–627.

Doll R, J Buch (1950): Hereditary factors in peptic ulcer. Ann Eugen (Lond) 15:135–146.

Doll R, TD Kellock (1951): The separate inheritance of gastric and duodenal ulcers. Ann Eugen (Lond) 16:231–240.

Doose H, H Gerken (1973): On the genetics of EEG anomalies in childhood. IV. Photoconvulsive reaction. Neuropädiatrie 4:162–171.

Dorus E, W Dorus, MA Telfer, S Litwin, CE Richardson (1976): Height and personality characteristics of 47,XYY males in a sample of tall non-institutionalized males. Br J Psychiatry 129:564–573.

Dudgeon JA (1975): Congenital rubella. J Pediatr 87:1078–1086.

Dutrillaux B, MO Rethoré (1975): Analyse du caryotype de Pan paniscus. Comparaison avec les autres Pongidae et l'Homme, Humangenetic 28:113–119.

Eberhard G (1968): Peptic ulcer in twins. A study in personality, heredity and environment. Acta Psychiatrica Scand, Suppl 205.

Edwards JH (1961): The syndrome of sex-linked hydrocephalus. Arch Dis Child 36:481–493.

Edwards JH (1973): Genetic counseling in cystic fibrosis. Lancet 2:919.

Ehrenberg L, GV Ehrenstein, A Hedgran (1957): Gonad temperature and spontaneous mutation rate in man. Nature 180:1433–1434.

Eldridge R (1974): Huntington's disease: some prefer not to know. "Medical World News." New York: McGraw Hill.

Elston RC, KK Namboodiri, CJ Glueck, R Fallat, R Tsang, V Leuba (1975): Study of the genetic transmission of hypercholesterolemia and hypertriglyceridemia in a 195 member kindred. Ann Hum Genet (Lond)39:67–87.

Epstein CJ, RP Erickson, BD Hall, MS Golbus (1975): The center-satellite system for the wide-scale distribution of genetic counseling services. Am J Hum Genet 27:322–332.

Erlenmeyer-Kimling L (1976): Schizophrenia: A bag of dilemmas. Soc Biol 23:123–134.

Evans HJ (1977): Chromosome anomalies among livebirths. J Med Genet 14:309–312.

Evers-Kiebooms G, H van den Berghe (1979): Impact of genetic counseling. Clin Genet 15:465–474.

Falconer DS (1965): The inheritance of liability to certain diseases, estimated from the incidence among relatives. Ann Hum Genet (Lond)29:51–76.

Falconer DS, LJP Duncan, C Smith (1971): A statistical and genetical study of diabetes. I. Prevalence and morbidity. Ann Hum Genet (Lond)34:347–369.

Farrow MG, RC Juberg (1969): Genetics and laws prohibiting marriage in the United States. JAMA 209:534–538.

Feldman MW, RC Lewontin (1975): The heritability hang-up. Sci 190:1163–1168.

Firshein SI, LW Hoyer, J Lazarchick, BG Forget, JC Hobbins, LP Clyne, PA Pitlick, WA Muir, IR Merkatz, MJ Mahoney (1979): Prenatal diagnosis of classic hemophilia. N Engl J Med 300:937–941.

Fogh-Anderson P (1942): "Inheritance of harelip and cleft palate. Copenhagen: A Busck.

Frantzen E, M Lennox-Buchthal, A Nygaard, J Stene (1970): A genetic study of febrile convulsions. Neurology 20:909–917.

Fraser FC (1970): A review: The genetics of cleft lip and cleft palate. Am J Hum Genet 22:336–352.

Fraser FC (1974): Genetic Counseling. Am J Hum Genet 26:636–659.

Fraser FC, CJ Biddle (1976): Estimating the risks for offspring of first-cousin matings: An approach. Am J Hum Genet 28:522–526.

Fraser FC, ADW Hunter (1975): Etiological relationships between categories of congenital heart malformations. Abstract 27th Annual Meeting of the American Society of Human Genetics. p 36A.

Friedrich U, T Lyngbye, J Øster (1977): Use of banding techniques for zygosity diagnosis in twins. Acta Genet Med Gemellol 26:89–91.

Fuchs F, J Philip (1963): Mulighed for antenatal and andersøgelse of fosterets kromosomer. Nord Med 9:69.

Gabbay KH, K DeLuca JN Fisher Jr, ME Mako, AH Rubenstein (1976): Familial hyperproinsulinemia. N Engl J Med 294:911–915.

Gajdusek DC, CJ Gibbs Jr, M Alpers (1967): Transmission and passage of experimental "kuru" to chimpanzees. Science 155:212–214.

Galjaard H (1976): European experience with prenatal diagnosis of congenital disease: a survey of 6,121 cases. Cytogenet Cell Genet 16:453–467.

Gardner RJM, C Alexander, AMO Veal (1974): Spina bifida occulta in the parents of offspring with neural tube defects. J Génét Hum 22:389–395.

Gemzell C, P Roos (1966): Pregnancies following treatment with special reference to the problem of multiple births. Am J Obstet Gynecol 94:490–496.

Gerken H, H Doose (1973): On the genetics of EEG anomalies in childhood. III Spikes and waves. Neuropädiatrie 4:88–97.

Gerrard JW, DC Rao, NE Morton (1978): A genetic study of immunoglobulin E. Am J Hum Genet 30:46–58.

Gershon ES, WE Bunney Jr (1977): The question of X-linkage in bipolar manic depressive illness. J Psychiatric Res 13:99–117.

Goad WB, A Robinson, TT Puck (1976): Incidence of aneuploidy in a human population. Am J Hum Genet 28:62–68.

230 / Literature Cited

Golbus MS, WD Loughman, CJ Epstein G Halbasch, JD Stephens, BD Hall (1979): Prenatal genetic diagnosis in 3,000 amniocenteses. N Engl J Med 300:157–163.

Goldstein JL, JJ Albers, HG Schrott, WR Hazzard, EL Bierman, AG Motulsky (1974): Plasma lipid levels and coronary heart disease in adult relatives of newborns with normal and elevated cord blood lipids. Am J Hum Genet 26:727–735.

Goldstein JL, MS Brown (1979): The LDL receptor locus and the genetics of familial hyper-cholesterolemia. Ann Rev Genet 13:259–289.

Goodman HO, SC Reed (1952): Heredity of fibrosis of the pancreas. Possible mutation rate of the gene. Am J Hum Genet 4:59–71.

Goodman MJ, CS Chung, F Gilbert Jr (1974): Racial variation in diabetes mellitus in Japanese and Caucasians living in Hawaii. J Med Genet 11:328–334.

Goodman MJ, CS Chung (1975): Diabetes mellitus: discrimination between single locus and multifactorial models of inheritance. Clin Genet 8:66–74.

Goodman RM, RJ Gorlin (1977): Atlas of the face in genetic disorders. Second ed, St. Louis: C V Mosby Co, p 542.

Gorlin RJ, J Cervenka, S Pruzansky (1971): Facial clefting and its syndromes. Birth Defects: Original Articles Series. The National Foundation – March of Dimes, New York: Stratton Intercontinental Medical Book Corporation. 7:3–49.

Gottesman II (1963): Heritability of personality: a demonstration. Psychol Monogr 77 (all of no. 572).

Gottesman II, J Shields (1966): Contributions of twin studies to perspectives on schizophrenia. Prog Exp Personality Res 3:1–84.

Gottesman II, J Shields (1967): A polygenic theory of schizophrenia. Proc Natl Acad Sci USA 58:199–205.

Grahn D (1972): Genetic effects of low level irradiation. Biosci 22:532–540.

Greer HS, S Lal, SC Lewis, E Belsey, RW Beard (1976): Psychosocial consequences of thera-peutic abortion. Br J Psychiatry 128:74–79.

Grundbacher FJ (1972): Immunoglobulins, secretor status, and the incidence of rheumatic fever and rheumatic heart disease. Hum Hered 25:399–404.

Hagnell O (1966): A prospective study of the incidence of mental disorder. Stockholm: Svenska Bokförlaget (Norstedts) Bonniers.

Halperin SL, DC Rao, NE Morton (1975): A twin study of intelligence in Russia. Behav Gen 5:83–86.

Hamerton JL, N Canning, M Ray, S Smith (1975): A cytogenetic survey of 14,069 newborn Infants. Clin Genet 8:223–243.

Hamilton M, GW Pickering, JA Fraser Roberts, GSC Sowry (1954): The aetiology of essential hypertension 4. The role of inheritance. Clin Sci 13:273–304.

Hanson JW, DW Smith (1975): The fetal hydantoin syndrome. J Pediatr 87:285–291.

Hanson JW, NC Myrianthopoulos, MAS Harvey, DW Smith (1976): Risks to the offspring of women treated with hydantoin anticonvulsants, with emphasis on the fetal hydantoin syn-drome. J Pediatr 89:662–668.

Harris H (1970): "The Principles of Human Biochemical Genetics'" Amsterdam: North Holland. pp 1–328.

Harrison GA (1973): Differences in human pigmentation: measurement, geographic variation, and causes. J Invest Derm 60:418–426.

Hartz A, E Giefer, AA Rimm (1977): Relative importance of the effect of family environ-ment and heredity on obesity. Ann Hum Genet (Lond)41:185–193.

Harvald B, M Hauge (1965): Hereditary factors elucidated by twin studies. US Public Health Service Publication No 1163. Washington, DC: US Government Printing Office.

Hassold TJ, A Matsuyama, IM Newlands, JS Matsuura, PA Jacobs, B Manuel, J Tsuei (1978): A cytogenetic study of spontaneous abortions in Hawaii. Ann Hum Genet (Lond) 41:443–454.

Hauge M, B Harvald, M Fisher, K Gotlieb-Jensen, N Juel-Nielsen, I Raebild, R Shapiro, T Videbeck (1968): The Danish twin register. Acta Genet Med Gemmellop 17:315–331.

Hauser WA, LT Kurland (1975): The epidemiology of epilepsy in Rochester, Minnesota, 1935 through 1967. Epilepsia 16:1–66.

Headings VE (1975): Alternative models of counseling for genetic disorders. Soc Biol 22:297–303.

Hecht F (1971): Cited in a talk given by D L VanDyke in Portland, Oregon, 1974.

Hedrick PW, E Murray (1978): Average heterozygosity revisited. Am J Hum Genet 30:377–382.

Heinonen OP, D Slone, RR Monson EB Hook, S Shapiro (1977): Cardiovascular birth defects and antenatal exposure to female sex hormones. N Engl J Med 296:67–70.

Henle G, U Koldovsky, P Koldovsky, W Henle, R Ackerman, G Haase (1975): Multiple sclerosis-associated agent: neutralization of the agent by human sera. Infect Immun 12:1367–1374.

Herskovits MJ (1930): The anthropometry of the American Negro. Columbia Univ Contr Anthropol II: 1–283.

Heston LL (1966): Psychiatric disorders in foster home reared children of schizophrenic mothers. Br J Psychiatry 112:819–825.

Higgins JV, EW Reed, SC Reed (1962): Intelligence and family size: a paradox resolved. Eugen Quart 9:84–90.

Hobbins JC, MJ Mahoney, LA Goldstein (1974): New method of intrauterine evaluation by the combined use of fetoscopy and ultra sound. Am J Obstet Gynecol 118:1069–1072.

Holmes LB, SG Driscoll, L Atkins (1976): Etiolgoic heterogeneity of neural-tube defects. N Engl J Med 294:365–369.

Holzmann H, R Anathakrishnan, B Morsches, H Walter, N Hoede, L Eckes (1973): Elektrophoretische utersuchungen der glucose-6-phosphate-dehydrogenase in erythrocyten von psoriatikern. Arch Derm Forsch 274:283–288.

Hook EB, KM Healy (1976): Consequences of a nationwide ban on spray adhesives alleged to be human teratogens and mutagens. Science 191:566–567.

Hook EB, GM Chambers (1977): Estimated rates of Down syndrome in live births by one year maternal intervals. In Bergsma D, Lowry RB (eds): "Numerical Taxonomy of Birth Defects and Polygenic Disorders." New York: Alan R Liss, No. 3A, 13:123–141.

Hook EB, A Lindsjö (1978): Down syndrome in live births by single year maternal age interval in a Swedish study. Am J Hum Genet 30:19–27.

Horn JM, J Loehlin, L Willerman (1979): Intellectual resemblance among adoptive and biological relatives: The Texas adoption project. Behav Genet 9:177–201.

Horton WA, RN Schimke, J Kennedy, A DeSmet (1979): Autosomal dominant inheritance of congenital dislocation of the hip. Am J Hum Genet 31:74A.

Howell JB, DE Anderson (1972): The nevoid basal cell carcinoma syndrome. In Andrade A, SL Gumport, GL Popkin, TD Rees (eds) "Cancer of the Skin." Philadelphia: W B Saunders.

Hsia YE, K Hirschhorn, RL Silverberg, L Godmilow (1979): "Counseling in Genetics." New York: Alan R. Liss, pp 1–347.

Huguenard JR, GE Sharples (1972): Incidence of congenital pyloric stenosis within sibships. J Pediatr 81:45–49.

Hunter H (1977): XYY males. Br J Psychiatr 131:468–477.

Idelberger K (1939): Die Ergebnisse der Zwillingsforschung beim angeborenen Klumpfuss. Verh Dtsch Orthop Ges 33:272–276.

Idelberger K (1951): Die erbpathologie der sogenannten angeborenen Huftverrenkung. Munich: Urban und Schwarzenberg.

Ishiguro T (1973): Alpha-fetoprotein in twin pregnancy. Lancet 2:1214.

Jacobs PA, RR Angell, IM Buchanan, TJ Hassold, AM Matsuyama, B Manuel (1978): The origin of human triploids. Ann Hum Genet (Lond)42:49–57.

Janerich DT, J Piper (1978): Shifting genetic patterns in anencephaly and spina bifida. J Med Genet 15:101–105.

Jones CW, IA Mastrangelo, HH Smith, HZ Liu, RA Meck (1976): Interkingdom fusion between human (HeLa) cells and tobacco (GGLL) protoplasts. Science 193:401–403.

Jones KL, DW Smith, CN Ulleland, AP Streissguth (1973): Pattern of malformations in offspring of chronic alcoholic mothers. Lancet 1.2: 1267–1271.

Jovanovic L, R Landesman, BB Saxena (1977): Screening for twin pregnancy Science 198:738.

Juberg RC, WJ Touchstone (1974): Congenital metatarsus varus in four generations. Clin Genet 5:127–132.

Kahn CB, JS Soeldner, RE Gleason, L Rojas, RA Camerini-Davalos, A Marble (1969): Clinical and chemical diabetes in offspring of diabetic couples. N Engl J Med 281:343–347.

Kajii T, N Niikawa, A Ferrier, H Takahara (1973): Trisomy in abortion material. Lancet, Nov 24, p 1214.

Kallmann FJ (1938): "The Genetics of Schizophrenia." New York: J J Augustin, 291 pp.

Karlsson JL (1966): "The Biological Basis of Schizophrenia." Springfield Ill: Charles C Thomas.

Kaufman S, NA Holtzman, S Milstein, et al (1975): Phenylketonuria due to deficiency of dihydropteridine reductase. N Engl J Med 293:785–790.

Kay DWK (1978): Assessment of familial risks in the functional psychoses and their application in genetic counseling. Br J Psychiatry 133:385–403.

Keith HM, RP Gage (1960): Neurological lesions in relation to asphyxia of the newbown and factors of pregnancy: long term follow-up. Pediatrics 26:616–622.

Kidd KK, MA Spence (1976): Genetic analysis of pyloric stenosis suggesting a specific maternal effect. J Med Genet 13:290–294.

King M-C, AC Wilson (1975): Evolution at two levels in humans and chimpanzees. Science 188:107–116.

Klawans HL Jr, GW Paulson, SP Ringel, A Barbeau (1972): Use of 1-Dopa in the detection of presymptomatic Huntington's chorea. N Engl J Med 286:1132–1334.

Klein D, D Wyss (1977): Retrospective and follow-up study of approximately 1000 genetic consultations. J Génét. Hum 25:47–57.

Kline J, ZA Stein, M Susser, D Warburton (1977): Smoking: a risk factor for spontaneous abortion. N Engl J Med 297:793–796.

Knudson AG, LC Strong, DE Anderson (1973): Heredity and cancer in man. Prog Med Genet 9:113–158.

Koch G (1976): Vaterschaft. Bibliographica genetica medica 7:1–220.

Koch G, G Schwanitz (1977): Die Proxis der Humangenetic. Ärztliche Praxis 44:2146–2151.

Koguchi H, K Tanaka (1976): Intensity of selection against cleft lip and/or cleft palate. Jpn J Hum Genet 20:321–336.

Konigsmark BW, RJ Gorlin (1976): "Genetic and Metabolic Deafness." Philadelphia: W B Saunders Company, pp 1–419.

Lalouel JM, NE Morton, CJ MacLean, J Jackson (1977): Recurrence risks in complex inheritance with special regard to pyloric stenosis. J Med Genet 14:408–414.

Langaney A, G Pison (1975): Probability of paternity: useless. Am J Genet 27:558–561.

Lam SK, GB Ong (1976): Duodenal ulcers: early and late onset. Gut 17:169–197.

Larson AT, SC Reed (1975): An investigation of the reactions and reproductive behaviors of persons subsequent to genetic counseling. Abstr Behav Genet 5:99–100.

Leck I (1976): Descriptive epidemiology of common malformations (excluding central nervous system defects). Br Med Bull 32:45–52.

Lejeune J, M Gautier, R Turpin (1959): Les chromosomes humains en culture de tissus. CR Acad Sci (Paris) 248:1721.

Lennox WG (1953): Significance of febrile convulsions. Pediatrics 11:341–357.

Lennox-Buchthal MA (1973): Febrile convulsions. Electroenceph Clin Neurophysiol (Suppl) 32:3–115.

Leonard CO, GA Chase, B Childs (1972): Genetic counseling: a consumers' view. N Engl J Med 287:433–439.

Levan G, F Mitelman (1975): Clustering of aberrations of specific chromosomes in human neoplasms. Hereditas 79:156–160.

Levine BB, RH Stember, M Fotino (1972): Ragweed hay fever: Genetic control and linkage HLA haplotypes. Science 178:1201–1203.

Levine P (1958): The influence of the ABO system on Rh hemolytic disease. Hum Biol 30:14–28.

Lewandowski RC Jr, JJ Yunis (1975): New chromosomal syndromes Am J Dis Child 129: 515–529.

Lewontin RC (1967): An estimate of average heterozygosity in man. Am J Hum Genet 19:681–685.

Li WH (1975): The first arrival time and mean age of a deleterious mutant gene in a finite population. Am J Hum Genet 27:274–286.

Lieber CS (1976): The metabolism of alcohol. Sci Am 234:25–33.

Loehlin JC, RC Nichols (1976): "Heredity, Environment, and Personality." Austin: University of Texas Press, 202 pp.

Loranger AW (1975): X-linkage and manic depressive illness. Br J Psychiatry 127:482–488.

Lowe CU, CD May, SC Reed (1949): Fibrosis of the pancreas in infants and children. Am J Dis Child 78:1–26.

Lowry RB, DHG Renwick (1969): Incidence of cleft lip and palate in British Columbia Indians. J Med Genet 6:67–69.

Lubs HA, F de la Cruz (1977): Genetic Counseling. New York: Raven Press, pp 1–598.

Lubs M-L E (1972): Empiric risks for genetic counseling in families with allergy. J Pediatr 80:26–31.

Lynch HT, H Guirgis, D Bergsman (1977): International Directory of Genetic Services. National Foundation – March of Dimes Birth Defects Series. New York: Alan R Liss, Inc.

MacIntyre MN (1973): Prenatal diagnosis, an essential to family planning in cases of genetic risk: In McCalister DV et al. (eds) "Readings in Family Planning." St Louis: Mosby.

Macklin MT (1960): Inheritance of cancer of the stomach and large intestine in man. J Natl Cancer Inst 24:551–571.

MacMahon B, T McKeown (1955): Infantile hypertrophic pyloric stenosis: data on 81 pairs of twins. Acta Genet Med Gemellol 4:320–329.

Marcusson J, E Moller, N Thyresson (1976): Penetration of HLA linked psoriasis predisposing gene(s): a family investigation. Acta Derm Venereol (Stock) 56:453–463.

Marx JL (1976): Atherosclerosis: the cholesterol connection. Science 194:711–714.

Maugh TH II (1977): The EAE model: a tentative connection to multiple sclerosis. Science 195:667–669, 768–771, 969–971.

McCabe MS, RC Fowler, RJ Cadoret, G Winokur (1972): Familial differences in schizophrenia with good and poor prognosis. Psychiatric Med 1:326–332.

McDermott WK, Dwuschle, J Adair, H Fulmer, B Loughlin (1960): Introducing modern medicine in a Navajo community. Science 131:280–287.

McKee WD (1966): The incidence and familial occurrence of allergy. J Allerg 38:226–235.

McKusick VA (1978): "Mendelian Inheritance in Man." Fifth ed. Baltimore: Johns Hopkins University Press, pp 1–975.

McKusick V (1975): Genetic counseling. Am J Hum Genet 27:240–242.

Melnick M, D Bixler, P Fogh-Anderson (1977): Cleft lip ± palate in Denmark 1941–1968. Am J Hum Genet 29:75A.

Mendelwicz J, JD Rainer (1974): Morbidity risk and genetic transmission in manic-depressive illness. Am J Hum Genet 26:692–701.

Meskin LH, RJ Gorlin, RJ Isaacson (1966): Cleft uvula – a microform of cleft palate. Acta Chir Plast 8:91–96.

Myers RH, DA Shafer (1977): Genetic studies of an intergeneric hybrid ape. Abstract of paper given at Seventh Annual Meeting of the Behavior Genetics Association. April 28, 1977, at Louisville.

Metrakos K, JD Metrakos (1961): Genetic and electroencephalographic studies in centrencephalic epilepsy. Neurology 11:474–483.

Miller DA (1977): Evolution of primate chromosomes. Science 198:1116–1124.

Miller EC, JA Miller (1971): The mutagenicity of chemical carcinogens: correlations, problems and interpretations. In Hollaender A (ed): "Chemical Mutagens." New York: Plenum Press, 1:83–119.

Milunsky A, GJ Annas (1976): "Genetics and the Law." New York: Plenum Press, pp 1–520.

Mitchell SC, SB Korones, HW Berendes (1971): Incidence of congenital heart disease in 54,033 births. Circulation 43:323–332.

Moertel CG, JA Bargen, MB Dockerty (1958): Multiple carcinomas of the large intestine. Gastroenterology 34:85–98.

Mourant AE, AC Kopec, K Domaniewska-Sobczak (1976): The Distribution of the Human Blood Groups and other Polymorphisms. New York: Oxford University Press, pp 1–1055.

Mulvihill J, R Miller, J Sraumeni (1977): "Genetics of Human Cancer." New York: Raven Press, pp 1–519.

Munk-Andersen E, J Weber, M Mikkelsen (1977): Amniocentesis in prenatal diagnosis. A controlled series of 78 cases. Clin Genet 11:18–24.

Munsinger H (1975): Children's resemblance to their biological and adopting parents in two ethnic groups. Behav Genet 5:239–254.

Müller HJ, HP Klinger, M Glasser (1975): Chromosome polymorphism in a human newborn population. II. Cytogenet Cell Genet 15:239–255.

Murphy EA, GA Chase (1975): "Principles of Genetic Counseling." Chicago: Year Book Medical Publishers, Inc.

Murray RF Jr (1978): Genetic counseling: boon or bane? In "The Tricentennial People." Ames, Iowa: Iowa State University Press, pp 29–47.

Myrianthopoulos NC (1970): Genetic aspects of multiple sclerosis. In "Handbook of Clinical Neurology." Amsterdam: North Holland Publishing Co, 9:85–106.

Myrianthopoulos NC, CS Chung (1974): Congenital malformations in singletons: epidemiologic survey. In: Birth Defects: original article series. The National Foundation – March of Dimes. New York: Stratton Intercontinental Medical Book Corp, 10:1–48.

Myrianthopoulos NC (1975): Congenital malformations in twins: epidemiologic survey. In Birth Defects: Original Article Series. The National Foundation – March of Dimes. New York: Stratton Intercontinental Medical Book Corp. No. 8, 11:1–39.

Myrianthopoulos NC, D Bergsma (eds) (1979): "Recent Advances in the Developmental Biology of Central Nervous System Malformations." New York: Alan R Liss, Inc. pp 1–130.

Nadler, HL (1969): Prenatal detection of genetic defects. J Pediatr 74:132–143.

Namboodiri KK, RC Elston, CJ Gueck, R Fallot, CR Buncher, R Tsang (1975): Bivariate analyses of cholesterol and triglyceride levels in families in which probands have type IIb lipoprotein phenotype. Am J Hum Genet 27:454–471.

NAS/NRC Committee (1976): Evaluation of testing for cystic fibrosis. J Pediatr 88:711–750.

Nathan PE, MM Andberg, PO Behan, VD Patch (1969): Thirty-two observers and one patient: a study of diagnostic reliability.J Clin Psychol 25:9–15.

Neel JV, SS Fajans, JW Conn, RT Davidson (1965): Diabetes mellitus. Washington, DC: US Government Printing Office, Public Health Service Publication #1163:105–132.

Neel JV (1971): The detection of increased mutation rates in human populations. Perspect Biol Med 14:522–537.

Neel JV, H Kato, WJ Schull (1974): Mortality in the children of atomic bomb survivors and controls. Genetics 76:311–326.

Nichols PL, VE Anderson (1973): Intellectual performance, race and socioeconomic status. Soc Biol 20:367–374.

Nichols WW (1966): The role of viruses in the etiology of chromosomal abnormalities. Am J Hum Genet 18:81–92.

Nicalsen SD (1978): Family studies of relation between Perthes disease and congenital dislocation of the hip. J Med Genet 15:296–299.

Niedemeyer E (1972): "The Generalized Epilepsies." Springfield: Charles C Thomas, pp 1–247.

Niswander JD, MV Barrow, GJ Bingle (1975): Congenital malformations in American Indian. Soc Biol 22:203–215.

Nora JJ (1971): Etiological factors in congenital heart diseases. Pediatr Clin N Am 18:1059–1074.

Nora JJ, JC Gilliland, RJ Sommerville, DG McNamara (1967): Congenital heart disease in twins. N Engl J Med 277:568–571.

Nora JJ, DG McNamara, F C Fraser (1967): Hereditary factors in atrial septal defect. Circulation 35:448–456.

Nora JJ, CW McGill, DG McNamara (1970): Empiric recurrence risks in common and uncommon congenital heart lesions. Teratology 3:325–329.

Nora JJ, FC Fraser (1974): "Medical Genetics: Principles and Practice." Philadelphia: Lea and Febiger, pp 1–399.

Nora JJ, AH Nora (1978): The evolution of specific genetic and environmental counseling in congenital heart diseases. Circulation 57:205–213.

Ödegård Ö (1972): The multifactorial theory of inheritance in predisposition to schizophrenia. In Kaplan AR (ed): "Genetics Factors in 'Schizophrenia'." Springfield, Ill: C C Thomas, pp 256–275.

Ohno S (1971): Genetic implication of karyological instability of malignant somatic cells. Physiol Rev 51:496–527.

Omenn GS (1978): Prenatal diagnosis of genetic disorders. Science 200:952–958.

Ouellette EM, HL Rosett NP Rosman, L Weiner (1977): Adverse effects on offspring of maternal alcohol abuse during pregnancy N Engl J Med 297:528–530.

Palmer RM (1964): Hereditary Clubfoot. Clin Orthop 33:138–146.

Parnell RW (1958): "Behavior and Physique." London: Edward Arnold.

Panny SR, AF Scott, JA Phillips, KD Smith, HH Kazazian, S Charache, CC Talbot (1979): Prenatal diagnosis of sickle cell disease by restriction endonuclease analysis: Limitations and advantages. Am J Hum Genet 31:58A.

Paty DW, J Furesz, D Boucher, CG Rand, CR Stiller (1976): Neurology 26:651–655.

Pauls DL (1979): Sex effect on the risk of mental retardation. Behav Genet 9:289–295. 134.

Pearn JH (1973): Patients' subjective interpretation of risks offered in genetic counselling. J Med Genet 10:129–134.

Pearson HA, LK Diamond (1971): The critically ill child; sickle cell disease crises and their management. Pediatr 48:629–635.

Pearson JS (1973): Family support and counselling in Huntington's disease. Committee to Combat Huntington's Disease, 250 West 57th St, New York, NY 10019.

Penry JE (ed) (1976): Epilepsy bibliography 1950–1975. Dept of Health, Education and Welfare Publication No. (NIH) 76–1186, pp 1–1860.

Polesky HF, HD Krause (1976): Blood typing in disputed paternity cases – capabilities of American laboratories. Family Law Quart 10:287–294.

Raine DN (1975): "The Treatment of Inherited Metabolic Disease." Lancaster, England: Medical Medical and Technical Publishing Co. Ltd, pp 294.

Rao DC, NE Morton, S Yee (1976): Resolution of cultural and biological inheritance by path analysis. Am J Hum Genet 28:228–242.

Rao CD, NE Morton, RC Elston, S Yee (1977): Causal analysis of academic performance. Behav Genet 7:147–159.

Record RG, JH Edwards (1958): Environmental influences related to the aetiology of congenital dislocation of the hip. Br J Prev Soc Med 17:8–22.

Record RG, T McKeown (1949): Congenital malformations of the central nervous system. I. Br J Prev Med 3:183–219.

Record RG, T McKeown (1950): Congenital malformations of the central nervous system. III. Br J Soc Med 4:217–220.

Reed EW, SC Reed (1965): "Mental Retardation: A Family Study." Philadelphia: W B Saunders Co, p 719.

Reed SC (1936): Harelip in the house mouse. Genetics 21:339–374.

Reed SC (1949): Counseling in Human Genetics. Dight Institute Bulletin number 6. University of Minnesota Press, pp 1–21.

Reed SC (1955): "Counseling in Medical Genetics." Philadelphia: W B Saunders Co, p 268.

Reed SC (1974): A short history of genetic counseling. Soc Biol 21:332–339.

Reed SC, VE Anderson (1973): Effects of changing sexuality on the gene pool. In de la Cruz FF, La Veck GD (eds): "Human Sexuality and the Mentally Retarded." New York: Brunner / Mazel.

Reed SC, EB Nordlie (1961): Genetic counseling: for children of mixed racial ancestry. Eugen Quart 8:157–163.

Reed SC, C Hartley, VE Anderson, VP Philips, NA Johnson (1973): "The Psychoses: Family Studies." Philadelphia: W B Saunders Co, p 578.

Reed TE (1970): Caucasian genes in American Negroes. Science 165:762–768.

Reich T, CR Cloninger, SB Guze (1975): The multifactorial model of disease transmission. Br J Psychiatry 127:1–10.

Reilly P (1977): "Genetics, Law and Social Policy." Cambridge Mass: Harvard University Press, pp 1–275.

Riccardi VM (1977): "The Genetic Approach to Human Disease." New York: Oxford Univ Press, pp 1–273.

Rimoin DL (1971): Inheritance in diabetes mellitus. Med Clin N Am 55:807–819.

Roberts DF, J Chavez, SDM Court (1970): The genetic component in child mortality. Arch Dis Child 45:33–38.

Roberts JAF (1962): Inherited diseases. In Burdette WJ (ed): "Methodology in Human Genetics." San Francisco: Holden-Day, Inc.

Robertson FW, AM Cumming (1979): Genetic and environmental variation in serum lipoproteins in relation to coronary heart disease. J Med Genet 16:85–100.

Robinson A, HA Lubs, D Bergsma (eds) (1979): Sex chromosome aneuploidy: Prospective studies on children. Birth Defects: Original Article Series. The National Foundation – March of Dimes. New York: Alan R Liss, Inc, 15:1–281.

Robinson J, K Tennes, A Robinson (1975): Amniocentesis: its impact on mothers and infants. A 1-year follow-up study. Clin Genet 8:97–106.

Robinson JT, RG Chitham, RM Greenwood, JW Taylor (1974): Chromosome aberrations and LSD. Br J Psychiatry 125:238–244.

Rodeck CH, S Campbell (1978): Early prenatal diagnosis of neural-tube defects by ultrasound-guided fetoscopy. Lancet 1.2. May 27, 1128–1129.

Rose DJ, PW Walsh, LL Leskovjan (1976): Nuclear power compared to what? Am Sci 64:291–299.

Rosenthal D (ed) (1963): "The Genain Quadruplets." New York: Basic Books.

Rosenthal D, PH Wender, SS Kety, F Schulsinger, J Welner, RO Rieder (1975): Parent-child relationships and psychopathological disorder in the child. Arch Gen Psychiatry 32:466–476.

Rotter JI, DL Rimoin, IM Samloff (1978): Genetic heterogeneity in diabetes mellitus and peptic ulcer. In Morton NE, Chung CS (eds): "Genetic Epidemiology." New York: Academic Press, pp 381–414.

Rowley PT, L Fisher, Mack Lipkin Jr (1979): Screening and genetic counseling for β-thalassemia trait in a population unselected for interest: effects on knowledge and mood. Am J Hum Genet 31:718–730.

Safra MJ, GP Oakley (1975): Association between cleft lip with or without cleft palate and prenatal exposure to diazepam. The Lancet ii:478–480.

Scarr-Salapatek S, RA Weinberg (1975): The war over race and IQ: When Black children grow up in white homes. Psychology Today 9:80–82.

Scarr-Salapatek S, RA Weinberg (1976): IQ test performance of black children adopted by white families. Am Psychol 31:726–739.

Schaap T, H Margalit, MM Cohen (1976): Cystic fibrosis-interaction between two loci? Excerpta medica Int Congr Seri No. 397:98–99.

Schaumann B, FD Peagler, RJ Gorlin (1970): Minor craniofacial anomalies among a Negro population. Oral Surg 29:566–575.

Seemanova E (1971): A study of children of incestuous matings. Hum Hered 21:108–128.

Shapiro S, D Slone et al (1976): Anticonvulsants and parental epilepsy in the development of birth defects. Lancet 1:272–275.

Shields J (1977): The major psychoses J Med Genet 14:327–329.

Shokeir MHK (1975): Investigation on Huntington's disease. III. Biochemical observations, a possibly predictive test? Clin Genet 7:354–360.

Shwachman H, M Kowalski, KT Khaw (1977): Cystic fibrosis: a new outlook. 70 patients above 25 years of age. Medicine 56:129–149.

Simpson NE (1964): Multifactorial inheritance: a possible hypothesis for diabetes. Diabetes 13:462–471.

Simpson NE (1969): Heritabilities of liability to diabetes when sex and age at onset are considered. Ann Hum Genet 32:283–303.

Sing CF, DC Shreffler, JV Neel, JA Napier (1971): Studies on genetic selection in a completely ascertained Caucasian population. II. Family analyses of 11 blood group systems. Am J Hum Genet 23:164–198.

Sing CF, MA Chamberlain, WD Block, S Feiler (1975): Analysis of genetic and environmental sources of variation in serum cholesterol in Tecumseh, Michigan. Am J Hum Genet 27:333–347.

Sing CF, JD Orr (1976): Analysis of genetic and environmental sources of variation in serum cholesterol in Tecumseh, Michigan. III. Identification of genetic effects using 12 polymorphic genetic blood marker systems. Am J Hum Genet 28:453–464.

Sing CF, JD Orr (1978): Analysis of genetic and environmental sources of variation in serum cholesterol in Tecumseh, Michigan. IV. Separation of polygene from common environment effects. Am J Hum Genet 30:491–504.

Slack J NC Nevin (1968): Hyperlipidaemic xanthomatosis. J Med Genet 5:4–8.

Smith C (1970): Heritability of liability and concordance in monozygous twins. Ann Hum Genet 34:85–91.

Smith C (1971): Discriminating between different modes of inheritance in genetic disease. Clin Genet 2:303–314.

Smith C (1971): Recurrence risks for multifactorial inheritance. Am J Hum Genet 23:578–588.

Smith C (1974): Concordance in twins: methods and interpretations. Am J Hum Genet 26:454–466.

Smith C (1976): Statistical resolution of genetic heterogeneity in familial disease. Ann Hum Genet (Lond) 39:281–291.

Smith C, DS Falconer, LJP Duncan (1972): A statistical and genetical study of diabetes II. Heritability of liability. Ann Hum Genet (Lond)35:281–299.

Smith JM (1974): Incidence of atopic disease. Med Clin N Am 58:1:3–24.

Smith SM, LS Penrose (1955): Monozygotic and dizygotic twin diagnosis. Ann Hum Genet (Lond) 19:273–289.

Speed RM, AW Johnston, HJ Evans (1976): Chromosome survey of total population of mentally subnormal in North-East of Scotland. J Med Genet 13:295–306.

Spence MA, J Westlake, K Lange; DP Gold (1976): Estimate of polygenic recurrence risk for cleft lip and palate. Hum Hered 26:327–336.

Stamatoyannopoulos G, S-H Chen, M Fukui (1975): Liver alcohol dehydrogenase in Japanese: high population frequency of atypical form and its possible role in alcohol sensitivity. Am J Hum Genet 27:789–796.

Steinberg AG, SW Becker, TB Fitzpatrick, RR Kierland (1951): Genetic and statistical study of psoriasis. Am J Hum Genet 3:267–281.

Steinberg AG, SW Becker, TB Fitzpatrick, RR Kierland (1952): A further note on the genetics of psoriasis. Am J Hum Genet 4:373–375.

Stephan U, EW Busch, H Kollberg, K Hellsing (1975): Cystic fibrosis detection by means of a test-strip. Pediatrics 55:35–38.

Stern C (1960): "Principles of Human Genetics." San Francisco: W H Freeman and Co. pp 1–753.

Stern C (1970): Model estimates of the number of gene pairs involved in pigmentation. Variability of the Negro-American. Hum Hered 20:165–168.

Stevenson AC, EA Cheeseman (1956): Heredity and rheumatic fever. Some later information about data collected in 1950–51. Ann Hum Genet 21:139–144.

Stotts EE (1978): The nurse's role during rubella exposure in pregnancy. Issues In Health Care of Women 1:40–44.

Stevensson AC, BCC Davison (1970): "Genetic Counselling." London: William Heinemann Ltd., pp 1–355.

Stine GJ (1977): "Biosocial Genetics." New York: MacMillan, pp 1–579. pp 1–185.

Sussman LN (1976): "Paternity Testing By Blood Grouping." Springfield, Ill: C C Thomas,

Sutton HE (1970): The haptoglobins. In Steinberg AG, Bearn AG (eds): "Chapter 6 in Progress in Medical Genetics." 7:163–216.

Svejgaard A, M Hauge, C Jersild, P Platz, LP Ryder, L Staub-Nielsen, M Thompson (1975): "The HLA System. Monographs in Human Genetics No. 7." Basel, New York: S Karger, pp 1–103.

Swanson DW, FA Dinello (1970): Severe obesity as a habituation syndrome. Archiv Gen Psychiatry 22(2):120–127.

Tanaka K, H Fujino, Y Fujita, H Tashiro, Y Sanui (1969): Cleft lip and palate: some evidences for the multifactorial trait and estimation of heritability based upon Japanese data. Jpn J Hum Genet 14:1–9.

Tattersal RB, DA Pyke (1972): Diabetes in identical twins. Lancet 2.2:1120–1125.

Tattersall RB, SS Fajans (1975): A difference between the inheritance of classical juvenile-onset and maturity onset type diabetes of young people. Diabetes 24:44–53.

Tattersall RB, SS Fajans (1975): Prevalence of diabetes and glucose intolerance in 199 offspring of 37 conjugal diabetic parents. Diabetes 24:452–462.

Taylor CC (1971): Marriages of twins to twins. Acta Genet Med Gemellol 20:96–113.

Ten Kate LP (1977): A method for analysing fertility of heterozygotes for autosomal recessive disorders, with special reference to cystic fibrosis, Tay-Sachs disease and phenylketonuria. Ann Hum Genet (Lond) 40:287–297.

Tennes K, M Puck, K Bryant, W Frankenburg, A Robinson (1975): A developmental study of girls with trisomy X. Am J Hum Genet 27:71–80.

Tew BJ, KM Laurence, H Payne, K Rawnsky (1977): Marital stability following the birth of a child with spina bifida. Br J Psychiatry 131:79–82.

Tijo JH, A Levan (1956): The chromosome number of man. Hereditas 42:1–6.

Tokuhata GK (1964): Familial factors in human lung cancer and smoking. Am J Pub Health 54:24–32.

Trimble BK, JH Doughty (1974): The amont of hereditary disease in human populations. Ann Hum Genet (Lond) 38:199–223.

Tsuboi T (1977): Genetic aspects of febrile convulsions. Hum Genet 38:169–173.

Tsboi T, S Endo (1977): Incidence of seizures and EEG abnormalities among offspring of epileptic patients. Hum Genet 36:173–189.

Uchida IA, CPV Lee, EM Byrnes (1975): Chromosome aberrations induced in vitro by low doses of radiation: nondisjunction in lymphocytes of young adults. Am J Hum Genet 27:419–429.

UK Collaborative Study (1977): Maternal serum alpha-fetoprotein measurement in antinatal screening for anencephaly and spina bifida in early pregnancy. Lancet 1.2:1323–1332.

Van Arsdel PP Jr, AG Motulsky (1959): Frequency and hereditability of asthma and allergic rhinitis in college students. Acta Genet Statist Med 9:101–114.

van den Berg BJ (1974): Studies on convulsive disorders in young children. IV. Incidence of convulsions among siblings. Devel Med Child Neurol 16:457–464.

VanDyke DL, CG Palmer, WE Nance, P-L Yu (1977): Chromosome polymorphism and twin zygosity. Am J Hum Genet 29:431–447.

Verma RS, H Dosik, HA Lubs Jr (1977): Demonstration of color and size polymorphisms in human acrocentric chromosomes by acridine orange reverse banding. J Hered 68:262–263.

Vesey JM (1947): Rheumatic fever in the Negro. US Naval Med Bull 47:805–809.

Vogel F (1970): The genetic basis of the normal human electroencephalogram (EEG). Humangenetik 10:91–114.

Walker S, J Andrews, NM Gregson, W Gault (1973): Three further cases of triploidy in man surviving to birth. J Med Genet 10:135–141.

Warburton D, IL Firschein, DA Miller, FE Warburton (1973): Karyotype of the chimpanzee, Pan troglodytes, based on measurements and banding pattern: comparison to the human karyotype. Cytogenet Cell Genet 12:453–461.

Warburton D, FC Fraser (1959): Genetic aspects of abortion. Clin Obstet Gynecol 2:22–35.

Watson CW, EM Marcus (1962): The genetics and clinical significance of photogenic cerebral electrical abnormalities, myoclonus and seizures. Trans Am Neurol Assoc 87:251–253.

Watson W, HM Cann, EM Farber, ML Nall (1972): The genetics of psoriasis. Arch Dermatol 105:197–207.

Weissman M (1979): Environmental factors in affective disorders. Hospital practice, vol 14, pp 103–109.

Went FW (1968): The size of man. Am Sci 56:400–413.

Wertelecki W, JM Graham Jr, FR Sergovich (1976): The clinical syndrome of triploidy. Obstet Gynecol 47:69–76.

Whitney G, GE McClearn, JC DeFries. Heritability of alcohol preference in laboratory mice and rats. J Hered 61:165–169.

Wilson MG, M Schweitzer (1954): Pattern of hereditary susceptibility in rheumatic fever. Circulation 10:699–704.

Winokur G, VL Tarra (1969): Possible role of X-linked dominant factor in manic-depressive disease. Dis Nerv Syst 30:89–94.

Witkin HA, SA Mednik, F Schulsinger, E Bakkestrøm, KO Christiansen, DR Goodenough, K Hirschorn, C Lundsteen, DR Owen, J Philip, DR Rubin, M Stocking (1976): Criminality in XYY and XXY men. Science 193:547–555.

Witkop CJ Jr, WC Quevedo Jr, TB Fitzpatrick (1978): Albinism. In Stanbury JB, JB Wyngaarden, DS Fredrickson (eds): "The Metabolic Basis of Inherited Disease." 4th ed, New York: McGraw-Hill, pp 283–316.

Wolff PH (1973): Vasomotor sensitivity to alcohol in diverse mongoloid populations. Am J Hum Genet 25:193–199.

Woolf CM (1958): A genetic study of carcinoma of the large intestine. Am J Hum Genet 10:42–47.

Woolf CM (1961): The incidence of cancer in the spouses of stomach cancer patients. Cancer 14:199–200.

Woolf CM (1971): Congenital cleft lip: a genetic study of 496 propositi. J Med Genet 8:65–83.

Woolf CM (1971): Congenital hip disease: Implications for genetic counseling. Soc Biol 18:10–17.

Woolf CM (1975): A genetic study of spina bifida cystica in Utah. Soc Biol 22:216–220.

Woolf CM, JH Koehn, SS Coleman (1968): Congenital hip disease in Utah: the influence of genetic and non-genetic factors. Am J Hum Genet 20:430–439.

Woolf CM, JA Turner (1969): Incidence of congenital malformation among live births in Salt Lake City, Utah, 1951–1961. Soc Biol 16:270–279.

Worrall EP, JP Moody, GJ Naylor (1975): Lithium in non-manic-depressives: antiaggressive effect and red blood cell lithium values. Br J Psychiatry 126:464–468.

Wright SW, NE Morton (1968): Genetic studies on cystic fibrosis in Hawaii. Am J Hum Genet 20:157–169.

Wynne-Davies R (1970): A family study of neonatal and late diagnosis congenital dislocation of the hip. J Med Genet 7:315–333.

Wynne-Davies R (1965): Family studies and aetiology of clubfoot. J Med Genet 2:227–232.

Yarbrough KM, PN Howard-Pebbles (1976): X-linked nonspecific mental retardation. Clin Genet 9:125–130.

Zackai EH, WJ Mellman, B Neiderer, JW Hanson (1975): The fetal trimethadione syndrome. J Pediatr 87:280–284.

Zonana J, DL Rimoin (1976): Inheritance of diabetes mellitus. N Engl J Med 295:603–605.

Zubin J, and professional staff (1974): The diagnosis and psychopathology of schizophrenia in New York and London. Schizophr Bull 11:80–102.

Index

Abnormal hemoglobins, 82—85
A-B-O and the Rhesus system,
    79—82
Affinous marriages, laws about, 76
Albinism
    from first cousin unions, 77
    Hardy-Weinberg law, 69—70
    varieties, 31
Alcaptonuria, from first cousin
    unions, 77
Alcohol
    death from, 207
    dependancy, 207
    heritabilities in mice, 207
    racial differences, 207
Allergies
    and HLA, 115
    and IgE, 115
    and social class, 114
Alpha-fetoprotein
    false positives, 124
    in mother's serum, 125
Amniocentesis
    description, 47
    disease monitored, 47
    Down syndrome, 49
    frequency of abnormalities, 51
    safety, 50—51
    selection of patients, 50
    sonic scanning, 50
    termination of pregnancy, 53
Aneuploidy, 34—42
Anoxia and effects upon IQ, 198
Ape hybrids, 5
Arnold-Chiari malformations, 121
Asbestos as a carcinogen, 205

Atherosclerosis
    cigarette smoking, 135
    correlation between sibs, 135
    hyperlipidemia, 134
    racial difference, 133
    sex ratio, 133
    twin concordance, 134
    xanthomata, 134
Atomic bombing of Japan, 198
Autosomal dominant inheritance, 62
Autosomal linkage, 67
Autosomal recessive inheritance,
    63—64

"Beware of the Biases," 68
Blindness, types, 217
Body types, Davenport's index,
    181
Breast cancer
    frequency in relatives, 117
    twin concordance, 117

Campodactyly, burden of, 232
Cancers
    of the large intestine, 118
    of the lung, 119-120
    of the skin, 119
    of the stomach, 118
Carcinogens and mutagens, 204
Centrencephalic epilepsy
    age of onset, 150
    EEG findings, 150
    frequency, 150
Chromosome polymorphisms in
    paternity determinations, 91
Chromosome translocations, 42—44

Clefts of lip and palate
  heritabilities, 138
  other defects, 136
  reproductive fitness, 140
  twin concordance, 136-139
  valium, 142
  Van Der Woude syndrome, 136
Cleft uvula, racial differences, 139
Clubfoot
  frequency, 142
  heritability, 142
  in Maoris, 141
  racial differences, 142-143
  twin concordance, 142
Concordance rates in twins, 100
Congenital heart defects
  Down syndrome, 127-128
  associations, 129
  recurrence risks, 128
  twin data, 127
Congenital hip disease
  birth order effect, 143
  breech malposition, 143
  Perthes disease, 147
  racial differences, 145
  seasonal effect, 144
  sex ratio, 144
  twin concordance, 144
  X-ray effects, 146
Consanguinity
  first cousin marriages, 75
  social and biological
    consequences, 75
  laws about, 76
"Constitutional" traits, 111–112
Continuous variations, 105
Costs of services, 14
Cystic fibrosis
  frequency, 93, 97
  parents organizations, 96
  selective advantage, 98
  sterility, 98

Davenport, C. B., 1
Deafness, types, 217

Diabetes mellitus
  age of onset, 153
  frequency, 153, 154
  heritabilities, 155
  heterogeneity, 156
  HLA association, 158
  racial differences, 157
  twin concordance, 156
Diagnosis
  importance of, 29
  Down syndrome, 30
  Ehlers-Danlos varieties, 31
  Huntington disease, 30
Dice, L. R., 1
Dight, C. F., 2
Dight Institute, founding of, 1
Discontinuous variations, 105
Discriminant functions, 170
Distinction between multigenic and
    Mendelian inheritance, 106
Down syndrome, diagnosis, 33–34,
  45
Down syndrome and myelogenous
  leukemia, 117
Duchenne type muscular
  dystrophy, 217
Duodenal ulcer
  age of onset, 159
  blood group associations, 160
  environmental factors, 160
  frequency, 158-159
  twin concordance, 161
Dwarfism, 29

Eating habits, 173-174
Essential hypertension
  clinical state, 136
  continuous variation, 137
Exceptional children, 181

Fabry's disease, treatment, 58
Febrile convulsions
  age of onset, 148
  EEG findings, 149
  frequency, 149

racial differences, 149
twin concordance, 149
Fetal alcohol syndrome,
    frequency, 221
Fetal hydantoin syndrome, in
    seizure cases, 152
Fetoscopy, 23, 51
Frequency of carriers in the
    population, 71

Galton's law, 172
Genetic counseling
    boon or bane, 23
    definition, 3, 9
    ethics, 12, 18–23
    eugenic or dysgenic, 11–12
    function, 10
    goals, 16
    results, 17–18
    size of demand, 24–26
    steps involved, 12–16
    term, 1
Genetic counselors
    accreditation, 12–13
    training program, 12–13
Genetic screening, 19, 20, 27

Haptoglobins, 86
HLA in paternity suits, 90–91
HLA system, 85–86
Height and weight, correlation
    of, 181
Hemizygote, illustration of, 216
Heritabilities
    calculation of, 106–108
    misuse of, 107, 170
    and twin concordance, 107
High voltage power lines, 201
Homocystinuria, 56
Homologies of human and
    chimpanzee chromosomes, 6
Human-tobacco cell hybrids, 5
Huntington disease, 214
Huntington disease, difficulties in
    counseling, 215

Ichthyosis, from first cousin
    unions, 77
Illegitimate births, statistics, 89
Incest
    not consanguineous, 74
    results of consanguineous, 74
Intelligence
    adoption studies, 179
    gifted children, 179
    heritability, 177
    number of gene loci involved, 177
    prewar Russia, 177
    retardation, 179
    Russian twins, 177

Klinefelter syndrome, 36

LSD, social damage, 203
Lymphocytes-leucocytes, 85–86

Malaria and sickle cell gene, 197
Maple syrup urine disease, 57
Marie's cerebellar ataxia, difficulties
    in counseling, 214
Marriages of twins to twins, 102
Mary Lyon effect, 41
Mental disorders
    disagreement between diagnoses,
        187-190
    diagnosis of proband and
        relative, 189
    heritabilities, 191, 195
    taxonomic problems, 187-188
    United Kingdon Cross-National
        Project, 188
    X-linked dominant gene, 194
Mental retardation
    adoption problems, 183-184
    chromosomal causes, 181-182
    "familial" type, 182
    Galton's law, 183
    "normalization," 184
    reproductive performance, 185
    twin concordance, 182
    X-linked, 182

Microcephaly, from first cousin
unions, 77
Minnesota Human Genetics
League, 3
Mistaken zygosity in twins, 102–103
Modifications of Mendelian ratios,
68
Multiple sclerosis
geographic differences, 210
HLA relationship, 211
viral agent, 211

Neural tube defects
frequencies, 123-124
geographic variation, 121
racial differences, 123
sex ratios, 122
twin data, 123
Nondisjunction, induced by
radiation, 199
Normal curve, properties of, 169

Obesity, 174

Paternity suits, 89
Percentage of Heterozygous gene
loci, 72
Pernicious anemia, 86–87
Personality traits in twins, 103–104
Phenylketonuria
from first cousin unions, 77
IQ of patients, 55–56
Philadelphia chromosome, 46
Photogenic seizures
age of onset, 150
EEG findings, 150
sex differences, 150
Polyploidy, 34
Psoriasis
age of onset, 162
frequency of, 162
and G-6PD, 162
heritability, 162
and HLA-antigens, 171
treatment, 171
twin concordance, 163

Psychoses
adopted children of parents, 191
cumulative risks for any person,
193
Genain quadruplets, 194
lithium, 194
social class, 192
twin concordance, 165
Pyloric stenosis
in first born children, 165
frequency, 164
maternal effect, 167
twin concordance, 165
Quasi-continuous variations, 105

Rabies, interest in, 222
Rare dominants and the
mutation rate, 214
Relation of temperature to
mutation rate, 206
Rheumatic heart disease
frequencies, 131
geographic variation, 131
racial differences, 132
streptococci, 131
twin concordance, 132
Risk and burden, 16
Risk factors, explaining them, 10
Rockefeller Foundation, support of
Dight program, 2
Role of physician in adoption
procedures, 175
Rubella
damage from, 208
vaccinations, 208

Sacral spot, 176
Sample size needed to detect new
mutations, 200
Schizoidia, other mental disorders,
189, 192
Scott Hamilton and Shwachman's
disease, 96
Sex ratios for quasi-continuous
traits, 108

Sickle cell anemia
   laws, 82–84
   screening for, 82–84
Sickle cell gene, advantage
   in heterozygotes, 197
Sjögren-Larsson syndrome, 216
Skin color, number of gene loci
   involved, 175
Slow acting viruses, 209–210
Somatotyping, 173
Spina bifida cystica, heritability, 123
Spontaneous abortions due to
   chromosome anomalies, 41–42
State Board of Health, human
   genetics unit, 2
Stomach ulcer, in sibs, 161
Survivors of a nuclear war, 199

Tay-Sachs screening, 71
Thalidomide and phocomelia, 205
Turner syndrome, 35
Twins

chorionic gonadotrophic levels
   in mothers, 101
   frequency in U.S.A., 99
   genetics of, 101
   neonatal deaths, 101
   tests for zygosity, 100
Two-tone babies, 176

Ultrasound-guided fetoscopy, 125

Vitamin-D-dependent rickets, 59

Wetzel grid, 173
Wilson's disease, 57–58
Wilson's disease, from first
   cousin unions, 77

Xerodermapigmentosa, from first
   cousin unions, 77
X-linked inheritance, 65–66
XYY males, attitude of parents,
   37–41